Delia Koo

Elements of Optimization

Springer Science+
Business Media, LLC

With Applications in Economics and Business

Delia Koo
Eastern Michigan University
Ypsilanti, MI 48197
USA

AMS Subject Classifications: 49-xx, 90Cxx

Library of Congress Cataloging in Publication Data

Koo, Delia.
  Elements of optimization, with applications in
economics and business.

  (Heidelberg science library)
  Bibliography: p.
  Includes index.
  1. Mathematical optimization. 2. Economics, Mathematical. 3. Business
mathematics. I. Title. II. Series.
QA402.5.K66   519.4   66-7629

© 1977 by Springer Science+Business Media New York
Originally published by Springer-Verlag New York in 1977

9 8 7 6 5 4 3 2 1

ISBN 978-0-387-90263-0   ISBN 978-1-4612-6358-6 (eBook)
DOI 10.1007/978-1-4612-6358-6

Preface

This book attempts to present the concepts which underlie the various optimization procedures which are commonly used. It is written primarily for those scientists such as economists, operations researchers, and engineers whose main tools of analysis involve optimization techniques and who possess a (not very sharp) knowledge of one or one-and-a-half year's calculus through partial differentiation and Taylor's theorem and some acquaintance with elementary vector and matrix terminology. Such a scientist is frequently confronted with expressions such as Lagrange multipliers, first- and second-order conditions, linear programming and activity analysis, duality, the Kuhn–Tucker conditions, and, more recently, dynamic programming and optimal control. He or she uses or needs to use these optimization techniques, and would like to feel more comfortable with them through better understanding of their underlying mathematical concepts, but has no immediate use for a formal theorem-proof treatment which quickly abstracts to a general case of n variables and uses a style and terminology that are discouraging to people who are not mathematics majors. The emphasis of this book is on clarity and plausibility. Through examples which are worked out step by step in detail, I hope to illustrate some tools which will be useful to scientists when they apply optimization techniques to their problems.

Most of the chapters may be read independently of each other—with the exception of Chapter 6, which depends on Chapter 5. For instance, the reader will find little or no difficulty in reading Chapter 8 without having read the previous chapters.

I am indebted to Alan Baquet, Luke Chan, David Cheng, Norman Obst, W. Allen Spivey, M. B. Suryanarayana, and P. K. Wong for having read all or portions of the manuscript and given me comments. Norman Obst has been most generous with his time and helpful suggestions and deserves special thanks. Anthony Koo initiated the idea of writing this book and supplied me with all the economic literature with which I was not familiar, and I thank him for his consideration and restraint in demanding my time during the course of writing. I am most grateful for the generous support of the Mathematics Department at Michigan State University, where I spent my sabbatical year to finish this book, and for the very kind and able assistance of Glendora Milligan, Mary Reynolds, and Leota Steadman.

Delia Koo

Contents

Notation

Throughout this book we use lower case boldface letters such as $\mathbf{x}, \mathbf{y}, \mathbf{u}$ to denote vectors. Column vectors are normally used, but, to save space, they are often written as transposes of row vectors, such as $\mathbf{x} = (x_1, \ldots, x_n)^{\mathrm{T}}$ or $\mathbf{x}^{\mathrm{T}} = (x_1, \ldots, x_n)$, where T means transpose.

Matrices are denoted by capital letters such as A, B, C, and we sometimes write

$$A = [a_{ij}].$$

The determinant of a matrix A is written as $\det[A]$ or $|A|$.

The derivative of a function $f(x)$ evaluated at $x = x_0$ is written

$$\frac{df}{dx}\bigg|_{x = x_0} \quad \text{or} \quad df(x_0)/dx.$$

Subscripts and superscripts are explained at the places where they appear. In Chapters 2 and 3, subscripts such as f_1 or f_2 are used to refer to partial derivatives. Thus $f_1 = \partial f(x_1, x_2)/\partial x_1$, $f_{11} = \partial^2 f(x_1, x_2)/\partial x_1^2$, $f_{12} = \partial^2 f(x_1, x_2)/\partial x_1 \partial x_2$. Subscripts, however, are also used in other places to label variables, *etc.*, but they are clearly specified. Superscripts are generally used to label constraints. Again, they are explained in the text.

When there may be doubt as to where a theorem or proof or example ends, the symbol $\square$ is used to indicate the end.

The formulations, tables, and figures are numbered according to chapter and section. Thus (2.3.1) means the first equation in Section 3 of Chapter 2. Table (5.3.1) means the first table in Section 3 of Chapter 5.

Some commonly used symbols are:

$\in$: is an element of;

$\sum_{i=1}^{n} a_i = a_1 + a_2 + \ldots + a_n$;

$\nabla f(x)$ is the gradient vector $(\partial f/\partial x_1, \partial f/\partial x_2, \ldots, \partial f/\partial x_n)^{\mathrm{T}}$;

$[a, b]$ is a closed interval. For example, $t \in [t_0, t_f]$ means $t_0 \leqslant t \leqslant t_f$.

1 Extrema of a Function of One Variable

1.1 Extreme points

Let $f(x)$ be a function of a real variable x. $f(x)$ is said to attain a *strict* or *strong local maximum* (or *minimum*) *value* at $x = x_0$ if $f(x_0)$ is greater (or less) than $f(x)$ for all values of x in the given domain which are in a neighborhood of x_0. This neighborhood may be as small as one chooses, *i.e.*, one may choose values of x as close to x_0 as one wishes.

$f(x)$ is said to attain a *strict* or *strong global maximum* (or *minimum*) at $x = x_0$ if $f(x_0)$ is greater (or less) than $f(x)$ for all values of x in the domain given.

$f(x)$ attains a *weak* (*local* or *global*) *maximum* (*minimum*) if instead of the strict inequality $>$ (or $<$), we use $\geqslant$ (or $\leqslant$).

The point $(x_0, f(x_0))$ will be referred to as the local or *global maximum* (*minimum*) *point*. An *extreme point*, or *extremum*, denotes either a maximum or minimum point. It turns out that the discussions involving the maximum of a function are essentially similar to those involving the minimum, so that an extremal problem may denote either a maximum or minimum problem.

In Figure 1.1.1, $f(x)$ is defined for x in the interval $a < x < b$. The points

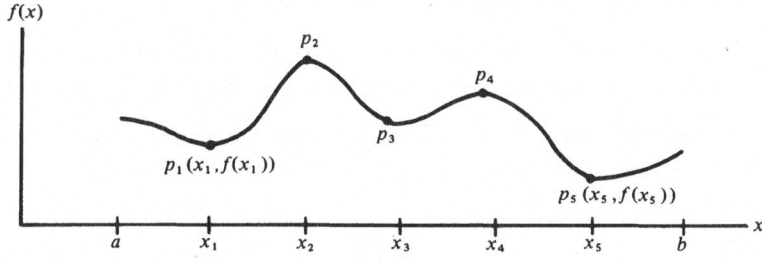

Figure 1.1.1.

p_1, p_3, and p_5 are local minima; p_2 and p_4 are local maxima. p_5 is the global minimum and p_2 the global maximum.

1.2 Necessary and sufficient conditions

In what follows we will be interested in the necessary and sufficient conditions for a local extremum. A *necessary condition* C_n for a statement S is a condition C_n that must be true in order for S to be true: To have S, we must have C_n. We may also express this by saying, "S only if C_n."

A *sufficient condition* C_s for a statement S is a condition such that S is true whenever C_s is true. We also say, "If C_s, then S," or, "S if C_s."

A necessary condition is not the same as a sufficient condition. If a condition C is both necessary and sufficient for a statement S, we call C the *necessary and sufficient condition* and sometimes say, "S if and only if C."

Sometimes people refer to a necessary condition as a first-order condition and a sufficient condition as a second-order condition. The reason for the terminology is that frequently a necessary condition for an extreme point involves the first-order derivatives, whereas a sufficient condition involves the second-order derivatives, as will be seen in the following sections. Care should be taken not to treat them as synonymous. In particular, the second-order conditions alone are not sufficient for an extremum.

Example 1. Let:

S be the statement that Jones lives in Michigan;
C_1 that Jones lives in the U.S.A.;
C_2 that Jones lives in Detroit;
C_3 that Jones lives in an American state which is considered the auto-
   mobile capital of the world.

C_1 is a necessary condition for S, but it is not sufficient. A person may live in the U.S.A. but not live in Michigan.

C_2 is a sufficient condition for S, but it is not necessary. A person who lives in Detroit lives in Michigan, but a resident of Michigan does not have to live in Detroit.

C_3 is a necessary and sufficient condition for S.

Example 2. Let $f(x)$ be a continuous function of x for $a < x < b$ and having a continuous second derivative at $x = x_0$, where $a < x_0 < b$, and let:

S be the statement that $f(x)$ attains a local minimum at $x = x_0$;
C_1 that $f'(x) = df(x)/dx = 0$ at $x = x_0$;
C_2 that $f'(x_0) = 0$ and $f''(x_0) > 0$.

It will be demonstrated later in this chapter that C_1 is a necessary condition for S, and C_2 a sufficient condition. However, C_1 is not

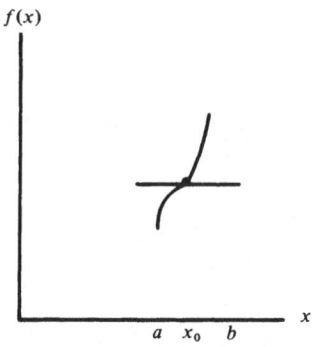

Figure 1.2.1.

sufficient, because $f'(x_0)$ may be zero without $(x_0, f(x_0))$ being an extremum. The point $(x_0, f(x_0))$ may be a point of inflection, as shown in Figure 1.2.1.

C_2, although sufficient, is not necessary. Consider the function $f(x) = x^4$. It attains a minimum at $x=0$, but $f''(0)$ is not positive; $f''(0) = 0$.

1.3 Taylor's formula for one variable

The familiar Taylor's formula will be used in the following discussion concerning the extremum of a continuous function of one variable. We shall not present any proof of Taylor's theorem, which is based on the Law of the Mean for derivatives, because it can be found in any introductory calculus text. Instead, we state:

Theorem 1.3.1 (Taylor's theorem). *Let a and b be numbers, $a < b$, let n be a positive integer, and let f be a function for which the nth derivative $f^{(n)}(x)$ exists for $a \leqslant x \leqslant b$. There exists a number ξ_n such that $a < \xi_n < b$ and*

$$f(b) = f(a) + f'(a)(b-a) + \frac{f''(a)}{2!}(b-a)^2 + \frac{f'''(a)}{3!}(b-a)^3 + \cdots$$

$$+ \frac{f^{(n-1)}(a)}{(n-1)!}(b-a)^{n-1} + \frac{f^{(n)}(\xi_n)}{n!}(b-a)^n \qquad \square$$

$$(1.3.1)$$

The number ξ_n depends on n, and, in general, one does not attempt to find the exact value of $f^{(n)}(\xi_n)$. It can be shown that $(f^{(n)}(\xi_n)/n!)(b-a)^n$ is small and that a bound can be established for it.

Example 1. Let $f(x) = e^x$, $a=0$, $b=1$, $n=4$. Then $f(b)=e$, $f(a)=e^0=1$, $f'(x)=f''(x)=f'''(x)=f^{(iv)}(x)=e^x$. Since $b-a=1$:

$$e = 1 + 1 + \frac{1}{2!} + \frac{1}{3!} + \frac{f^{(iv)}(\xi_4)}{4!} \quad \text{where } 0 < \xi_4 < 1; \qquad \left| \frac{f^{(iv)}(\xi_4)}{4!} \right| < \frac{e}{24} < \frac{1}{8}.$$

Example 2. $f(x) = \log x$, $a = 1$, $b = 1.5$, $n = 4$. $f'(x) = 1/x$, $f''(x) = -x^{-2}$, $f'''(x) = 2x^{-3}$, $f^{(iv)}(x) = -6x^{-4}$. Thus:

$$\log 1.5 = 0 + \frac{1}{2} - \frac{1}{2!}\left(\frac{1}{2}\right)^2 + \frac{2}{3!}\left(\frac{1}{2}\right)^3 - \frac{6}{4!}\left(\frac{1}{\xi_4}\right)^4\left(\frac{1}{2}\right)^4 \quad \text{where } 1 < \xi_4 < 1.5;$$

$$\left| -\frac{6}{4!}\left(\frac{1}{\xi_4}\right)^4\left(\frac{1}{2}\right)^4 \right| < \frac{1}{64}.$$

1.4 Necessary condition for an interior extremum

If a continuous function $f(x)$ is to have a local maximum at an interior point $x = x_0$, then it is obvious that $f(x_0) - f(x) > 0$ for all x in a neighborhood of x_0. Let us now confine ourselves to the first two derivatives of f, which are assumed to be continuous at x_0 and in a neighborhood around it, and rewrite Taylor's formula as

$$f(x) - f(x_0) = \left[f'(x_0) + \tfrac{1}{2} f''(\xi)(x - x_0) \right](x - x_0), \qquad (1.4.1)$$

where ξ lies between x and x_0. Since $f''(x)$ is continuous and the term $\tfrac{1}{2} f''(\xi)(x - x_0)$ is bounded, we can choose a suitable neighborhood around x_0 so that this term is smaller in absolute value than an arbitrary number. Suppose $f'(x_0) \neq 0$. Then we can confine ourselves to that neighborhood around x_0 so that the expression inside the brackets in (1.4.1) will have everywhere the same sign as $f'(x_0)$. Since there are points x in this neighborhood such that $x > x_0$ $(x - x_0 > 0)$ and $x < x_0$ $(x - x_0 < 0)$, it follows that in this neighborhood there are points x such that $f(x) - f(x_0) > 0$ and other points x such that $f(x) - f(x_0) < 0$. Therefore $f(x_0)$ cannot be a maximum in this neighborhood if $f'(x_0) \neq 0$.

A similar argument shows that $f(x_0)$ cannot be a local minimum if $f'(x_0) \neq 0$, and we have the theorem:

Theorem 1.4.1. *Let $f(x)$ be a continuous[1] function having a continuous second derivative (hence also a continuous first derivative) at and around the point x_0. A necessary condition for $f(x_0)$ to be a local interior extremum is that $f'(x_0) = 0$.*   □

All points $(x_0, f(x_0))$ where $f'(x_0) = 0$ are called *critical points* or *stationary points*.

---

[1]Actually, the word "continuous" is superfluous here. The reader may recall from the definition of the derivative that a necessary condition for a function to have a derivative at a certain point or in an interval is that it be continuous at that point or in that interval.

1.5 Sufficient condition for an extremum

Suppose now $f'(x_0)=0$ and $f(x)$ has a continuous third derivative. Then, taking $n=3$ and rewriting Formula (1.3.1), we have

$$f(x)-f(x_0)=\tfrac{1}{2}\left[f''(x_0)+\tfrac{1}{3}f'''(\xi)(x-x_0)\right](x-x_0)^2, \qquad (1.5.1)$$

where ξ lies between x and x_0. Again, since $f'''(\xi)(x-x_0)$ can be made arbitrarily small within a suitably chosen neighborhood of x_0, the expression within the brackets, and therefore the entire right hand side of (1.5.1), may bear the same sign as $f''(x_0)$ if $f''(x_0)\neq 0$. Then, in this neighborhood, $f(x)-f(x_0)>0$ if $f''(x_0)>0$, in which case $f(x_0)$ is a local minimum. And $f(x)-f(x_0)<0$ if $f''(x_0)<0$, in which case $f(x_0)$ is a local maximum. This proves the following theorem:

Theorem 1.5.1. *Let $f(x)$ be a continuous function with a continuous third derivative at and around x_0. A sufficient condition for $f(x_0)$ to be a local minimum (maximum) is that $f'(x_0)=0$ and that $f''(x_0)>0$ ($f''(x_0)<0$).* $\square$

Example 1.[2] Let $f(x)=x^3-3x^2$. Decide whether $f(1)$ is a local extremum.

Solution. $f'(x)=3x^2-6x=3x(x-2)$, so that $f'(1)=-3\neq 0$. Hence, $f(1)$ is not a local extremum.

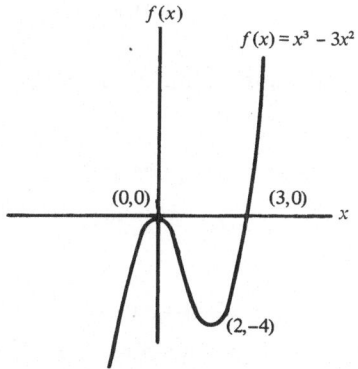

$f(x)$

$f(x)=x^3-3x^2$

(0,0)   (3,0)

x

(2,−4)

Figure 1.5.1.

Example 2. Find the extrema for the function $f(x)=x^3-3x^2$.

Solution. We equate $f'(x)$ to 0 and solve for x: $f'(x)=3x(x-2)=0$ when $x=0$ or $x=2$. Then we examine $f''(0)$ and $f''(2)$: $f''(x)=6x-6$; $f''(0)=-6$ <0, and $f''(2)=6>0$. Hence the given function has a maximum at $x=0$ and a minimum at $x=2$.

---

[2]For the following examples, let the domain of each function be the real line.

Example 3. In economics one is interested in maximizing the total profit $T = T(q)$ in the sale of some commodity (Q). Let q be the quantity of the commodity manufactured, $p = p(q)$ the sale price per unit, and $C = C(q)$ the total cost of production. Since total profit = total revenue (pq) — total cost, we have

$$T = pq - C.$$

In order that T be maximized, it is necessary that

$$\frac{dT}{dq} = \frac{d(pq)}{dq} - \frac{dC}{dq} = 0, \quad \text{i.e.,} \quad \frac{d(pq)}{dq} = \frac{dC}{dq};$$

that is, it is necessary for marginal revenue to be equal to marginal cost in order to achieve maximum profit.

Suppose that it is desired to find the most profitable level of production of a certain commodity given that

$$C = q^3 - 10q^2 + 17q + 66 \quad \text{and} \quad p = \$5.$$

Then $T = 5q - C = -q^3 + 10q^2 - 12q - 66$, and

$$\frac{dT}{dq} = -3q^2 + 20q - 12.$$

Equating dT/dq to 0 and solving for q, we get $q = 6$ or $\frac{2}{3}$. At $q = 6$,

$$\frac{d^2T}{dq^2} = -6q + 20 = -16,$$

which is negative. Thus, 6 units should be produced to achieve maximum profit.

Example 4. As another example, suppose that

$$C = 40q + 20{,}000$$

and

$$p = 160 - 0.01q,$$

where q is the number of units produced each week and the price and cost are measured in cents. Then:

$$\text{Total revenue} = pq = 160q - 0.01q^2;$$

$$\text{Marginal revenue} = \frac{d(pq)}{dq} = 160 - 0.02q;$$

$$\text{Marginal cost} = \frac{dC}{dq} = 40.$$

In order to achieve maximum profit,

$$\frac{d(pq)}{dq} = \frac{dC}{dq},$$

so that $160-0.02q=40$, *i.e.*, $q=6{,}000$ units per week.

$$\frac{d^2T}{dq^2}\bigg|_{q=6{,}000} = -0.02<0.$$

At that level of production,

$$p=160-60=100 \text{ cents or \$1 per unit}$$

and the total revenue will be

$$T=pq-C=100\cdot6000-(40\cdot6000+20{,}000)=\$3{,}400.$$

In applied problems such as this one, we frequently stop at the first derivative, although we should realize that the critical point does not guarantee an extremum. In this case it can be easily verified by simple calculation that $q=6{,}000$ does yield the most profitable level of production.

Example 5. In statistics, a familiar problem is to find the maximum likelihood estimator of a parameter of a probability distribution function. Suppose a random variable X has the Poisson distribution

$$f(x)=\frac{e^{-\lambda}\lambda^x}{x!} \qquad \text{for } x=0,1,2,3,\ldots, \text{ and } \lambda\geqslant 0.$$

To find the maximum likelihood estimator of the parameter λ, make n independent observations $x_1, x_2, \ldots, x_n$ which are identically distributed with

$$f(x_i)=\frac{e^{-\lambda}\lambda^{x_i}}{x_i!} \qquad i=1,2,\ldots,n.$$

Then form the likelihood function $L(\lambda)$, which is the joint distribution of the n X_i's:

$$L(\lambda)=g(x_1,x_2,\ldots,x_n;\lambda)=\prod_{i=1}^{n} f(x_i)=\frac{e^{-n\lambda}\lambda^{\sum_{i=1}^{n}x_i}}{\prod_{i} x_i!}$$

(see Footnote 3). The maximum likelihood estimator $\hat{\lambda}$ is obtained by finding that value of λ which maximizes $L(\lambda)$ or, equivalently, $\log L(\lambda)$. In this problem, it is simpler to find

$$\frac{d\log L(\lambda)}{d\lambda}=\frac{d}{d\lambda}(-n\lambda+(\Sigma x_i)\log\lambda-\log\Pi x_i!)=-n+\frac{\Sigma x_i}{\lambda}.$$

The necessary condition for an extremum requires that

$$-n+\frac{\Sigma x_i}{\lambda}=0.$$

---

[3] The notations $\Pi_{i=1}^{n}a_i$ and $\Sigma_{i=1}^{n}a_i$ are product and summation symbols. $\Pi_{i=1}^{n}a_i=a_1\cdot a_2\cdots\cdot a_n$ and $\Sigma_{i=1}^{n}a_i=a_1+a_2+\cdots+a_n$.

Solving, we get $\lambda = \Sigma x_i / n$. Either by direct substitution or by evaluating

$$\frac{d^2 \log L(\lambda)}{d\lambda^2} = \frac{-\Sigma x_i}{\lambda^2} < 0$$

we see that this value of λ maximizes $\log L(\lambda)$ as well as $L(\lambda)$, so that $\hat{\lambda} = \Sigma x_i / n$.

It is customary to let the symbol $\bar{x}$ stand for $\Sigma x_i / n$. Thus $\hat{\lambda} = \bar{x}$.

Example 6 (Baumol, Vickers). Consider a firm which wishes to minimize the cost of inventory investment, given a specified volume of sales of its commodity and the purchase price of its inventory. Let:

V = firm's annual volume of sales;
p = purchase price per unit of inventory;
q = quantity of each inventory purchase;
i = rate of interest on funds invested in inventory stocks;
a = fixed annual cost of purchasing inventory;
b = variable costs per inventory purchase.

The number of inventory purchases per year is V/q and the average inventory stock on hand is $q/2$ if we assume that inventory stocks are depleted at a uniform rate. The annual carrying cost of the inventory investment is

$$\frac{ipq}{2}.$$

This should be added to the annual purchase cost, which is $a + (bV/q)$, to yield the total annual cost of the inventory investment

$$c = c(q) = a + \frac{bV}{q} + \frac{ipq}{2}.$$

Differentiating c with respect to q and equating that to zero, we get

$$\frac{dc}{dq} = -\frac{bV}{q^2} + \frac{ip}{2} = 0,$$

giving

$$q = \sqrt{\frac{2bV}{ip}},$$

which is the optimal size of the purchase to give the minimum total cost. This is confirmed by the second derivative

$$\frac{d^2c}{dq^2} = \frac{2bV}{q^3} > 0.$$

This shows that the optimum size of each inventory purchase, hence also the average stock of inventory, is proportional to the square root of the firm's total volume of sales.

Suppose that the purchase price p is not a constant but is a function of q. For instance, there may be quantity discounts. Then

$$c = a + \frac{bV}{q} + \frac{ip(q)q}{2}$$

and

$$\frac{dc}{dq} = -\frac{bV}{q^2} + \frac{i}{2}\left(p + q\frac{dp}{dq}\right),$$

which, when equated to zero, now gives

$$q^2\left(p + q\frac{dp}{dq}\right) = \frac{2bV}{i},$$

so that

$$q = \sqrt{\frac{2bV}{i\left(p + q\dfrac{dp}{dq}\right)}}.$$

This shows that q is still proportional to the square root of the volume of sales but is inversely related to the size of $p + q(dp/dq)$. In general we would expect dp/dq to be negative, because of the possibility of quantity discounts. Thus, if the inventory purchase price declines sharply relative to the quantity purchases, the optimal quantity of inventory purchase becomes much larger.

Example 7 (Baumol). The following is an interesting example of how one utilizes the techniques for determining a maximum in order to prove the economic principle that, given the level of investment, a rise in the interest rate will reduce the optimal length of operation of a firm. Let:

$P =$ anticipated profit of the firm at the time the investment is made;
$I =$ the amount invested;
$t =$ length of time for which the investment runs;
$r =$ rate of interest.

Suppose the value V of the product is a function of I and t. The present discounted revenue, discounted at the interest rate r, is usually given to be Ve^{-rt}. Therefore

$$P = Ve^{-rt} - I = f(I,t)e^{-rt} - I.$$

In order to maximize P with respect to t, we set

$$\frac{\partial P}{\partial t} = \frac{\partial f(I,t)}{\partial t}e^{-rt} - rf(I,t)e^{-rt} = 0,$$

giving $f_t - rf = 0$ where $f_t = \partial f/\partial t$.

The second derivative gives

$$\frac{\partial^2 P}{\partial t^2} = f_{tt}e^{-rt} - rf_t e^{-rt} - re^{-rt}(f_t - rf)$$

$$= e^{-rt}(f_{tt} - rf_t) - re^{-rt} \cdot 0 < 0$$

where $f_{tt} = \partial^2 f/\partial t^2$.

Since e^{-rt} is never negative, this means that

$$f_{tt} - rf_t < 0.$$

If we take the differential of

$$f_t - rf = z = 0,$$

we have $dz = 0 = (f_{tt} - rf_t)dt - fdr$, so that

$$\frac{dt}{dr} = \frac{f}{f_{tt} - rf_t}.$$

For maximum profit, the denominator is negative. The numerator, f, is the value of the product, and is presumably positive. Therefore we have

$$\frac{dt}{dr} < 0,$$

which means that the length of time for which the investment runs will decrease if the interest rate increases.

1.6 Necessary and sufficient condition for an interior extremum

As we pointed out in Section 1.2, a necessary condition is not always sufficient and a sufficient condition is not always necessary. The following theorem gives a necessary and sufficient condition for a local extremum:

Theorem 1.6.1. *A necessary and sufficient condition for a function $f(x)$ which has continuous derivatives of all orders up to and including n at x_0 to have a local interior extremum at x_0 is that $f^{(k)}(x_0) = 0$, $k = 1, 2, \ldots, n-1$, and $f^{(n)}(x_0) \neq 0$, where n is an even integer greater than 1. $f(x_0)$ is a local maximum if $f^{(n)} < 0$, and a local minimum if $f^{(n)}(x_0) > 0$.*

Proof. The proof is almost identical to the considerations leading to Theorems 1.4.1 and 1.5.1. Let us look at the special case where $n = 3$, that is, the third derivative at x_0 is the first nonzero derivative. Then Taylor's formula can be written as

$$f(x) - f(x_0) = \frac{1}{3!}\left[f'''(x_0) + \frac{1}{4}f^{(iv)}(\xi)(x - x_0)\right](x - x_0)^3. \quad (1.6.1)$$

As before, the expression within the brackets may be made to bear the same sign as $f'''(x_0)$ within a suitably chosen neighborhood of x_0. But since $(x - x_0)^3$ will change sign for $x < x_0$ or $x > x_0$ within this neighborhood, $f(x) - f(x_0)$ will also change sign and $f(x_0)$ cannot be a local extremum.

The same argument applies for any n which is odd. Thus, if $f(x_0)$ is to be a local extremum, it is necessary that the first nonzero derivative at x_0 not be of odd order.

Now suppose n is even. Then Taylor's formula can be written as

$$f(x)-f(x_0)=\frac{1}{n!}\left[f^{(n)}(x_0)+\frac{1}{n+1}f^{(n+1)}(\xi)(x-x_0)\right](x-x_0)^n. \quad (1.6.2)$$

If $f^{(n)}(x_0)<0$, then the right-hand side of (1.6.2) is always negative, $f(x)-f(x_0)<0$, and $f(x_0)$ is a local maximum. If $f^{(n)}(x_0)>0$, then $f(x)-f(x_0)>0$ and $f(x_0)$ is a local minimum. □

Example. When $f(x)=x^4$, $f'(0)=f''(0)=f'''(0)=0$, and $f^{(4)}(0)=24$. Hence $f(x)$ attains a minimum at $x=0$. □

In Theorem 1.6.1, we can also add the statement: If n is an odd integer greater than 1, $f(x)$ has a horizontal inflection point at x_0.

1.7 Remarks

We should call attention to two assumptions in our discussion: (1) The derivatives of all orders needed are continuous and bounded in the neighborhood under consideration; (2) We have not considered the end points of a closed interval in the analysis of extremal points. The above theorems do not apply in the case of the end points. For example, if we wish to find the extrema of the function $f(x)=2x$ in the interval $1 \leqslant x \leqslant 2$, then $f(1)$ is the minimum and $f(2)$ the maximum, but $f'(x)$ is not zero at either point.

In what follows, we shall be examining only interior extremal points, unless we state otherwise. We will not emphasize the distinction between a strong or weak extremum, because much of the discussion (with possibly minor adjustments) applies to both types of extremum.

Exercises

1.1 Given the statements: I Smith is professor emeritus of ABC University; II Smith was professor at ABC; III Smith retired from teaching last year. Is any statement necessary or sufficient for any of the other two?

1.2 Let $f(x)=-|x|+4$ for $-4\leqslant x\leqslant 4$. Find the extreme points of the function. Does the derivative equal 0 at these points?

1.3 Determine the maximum profit and the corresponding price and quantity for a monopolist whose demand and cost functions are:

(a) $p=40-0.5q$, $\quad C=0.2q^3-6.5q^2+61.6q$;
(b) $p=30-0.1q$, $\quad C=0.3q^2+18q$;
(c) $p=15-q$, $\quad C=0.4q^3-8.2q^2+53.4q$.

1.4 Suppose the government imposes a tax of t dollars per unit of the commodity produced by the monopolist. This increases his cost by $t \cdot q$, where q is the quantity produced. Determine the maximum profit and the corresponding price and quantity when the demand and before tax cost functions are;

(a) $p = 40 - q$, $C = 3q$,   and $t = \$1$;
(b) $p = 30 - 0.45q$, $C = 6q$,   and $t = 0.05q$;
(c) $p = 10 - 0.2q$, $C = 1.72q^2 + 2q + 100$,   and $t = 0.08q$.

1.5 A random variable has the Poisson probability distribution with parameter λ. Five independent observations are made yielding $x_1 = 2.8$, $x_2 = 1.3$, $x_3 = 4.6$, $x_4 = 1.7$, $x_5 = 2.1$. Find the maximum likelihood estimator of λ.

2

Extrema of a Function of Two or More Variables (without Constraint)

2.1 Necessary condition

Let $f(x_1, x_2)$ be a function of two real variables x_1 and x_2 with continuous partial derivatives

$$f_1 = \frac{\partial f(x_1, x_2)}{\partial x_1} \quad \text{and} \quad f_2 = \frac{\partial f(x_1, x_2)}{\partial x_2}. \tag{2.1.1}$$

Suppose that $f(x_1, x_2)$ attains a local extremum at the point (a, b). Then intuitively it is clear that the function of a single variable $f(x_1, b)$ must attain an extremum at $x_1 = a$. From Section 1.4 it is necessary that $f_1 = 0$ at $x_1 = a$. Similarly, since the function $f(a, x_2)$ must also attain an extremum at $x_2 = b$, it is also necessary that $f_2 = 0$ at $x_2 = b$.

2.2 Taylor's formula for a function of two variables—necessary condition

One can arrive at the above necessary condition in a more rigorous fashion by using Taylor's formula for a function of two variables. This formula also enables one to obtain a sufficient condition for an extremum.

Let $f(x_1, x_2)$ be a function of x_1 and x_2 which has continuous partial derivatives of order n in the neighborhood of and at the point (a, b). Taylor's formula states that

$$
\begin{aligned}
f(x_1, x_2) &= f(a, b) + \left[f_1(a, b)(x_1 - a) + f_2(a, b)(x_2 - b) \right] \\
&\quad + \frac{1}{2} \left[f_{11}(\xi_1, \xi_2)(x_1 - a)^2 + 2f_{12}(\xi_1, \xi_2)(x_1 - a)(x_2 - b) + f_{22}(\xi_1, \xi_2)(x_2 - b)^2 \right] \\
&= f(a, b) + \left[f_1(a, b)(x_1 - a) + f_2(a, b)(x_2 - b) \right] \\
&\quad + \frac{1}{2!} \left[f_{11}(a, b)(x_1 - a)^2 + 2f_{12}(a, b)(x_1 - a)(x_2 - b) + f_{22}(a, b)(x_2 - b)^2 \right] \\
&\quad + \frac{1}{3!} \left[(x_1 - a)\frac{\partial}{\partial x_1} + (x_2 - b)\frac{\partial}{\partial x_2} \right]^3 f(a, b) + \cdots + R_n,
\end{aligned}
\tag{2.2.1}
$$

where

$$R_n = \frac{1}{n!}\left[(x_1-a)\frac{\partial}{\partial x_1}+(x_2-b)\frac{\partial}{\partial x_2}\right]^n f(\xi_1,\xi_2).$$

In (2.2.1),

$$\xi_1 = a+\theta(x_1-a) \quad \text{and} \quad \xi_2 = b+\theta(x_2-b), \quad \text{where } 0<\theta<1. \quad (2.2.2)$$

This merely requires the point (ξ_1,ξ_2) to lie on the line segment joining the points (x_1,x_2) and (a,b). Since (x_1,x_2) is a variable point, (ξ_1,ξ_2) will depend on (x_1,x_2).

We use the abbreviations $f_i(a,b)$ to mean $\partial f(x_1,x_2)/\partial x_i$ evaluated at (a,b) and $f_{ij}(\xi_1,\xi_2)$ to mean $\partial f(x_1,x_2)/\partial x_i \partial x_j$ evaluated at (ξ_1,ξ_2). And, similarly,

$$\left[(x_1-a)\frac{\partial}{\partial x_1}+(x_2-b)\frac{\partial}{\partial x_2}\right]^n f(\xi_1,\xi_2)=(x_1-a)^n\frac{\partial^n f}{\partial x_1^n}$$

$$+n(x_1-a)^{n-1}(x_2-b)\frac{\partial^n f}{\partial x_1^{n-1}\partial x_2}+\cdots+(x_2-b)^n\frac{\partial^n f}{\partial x_2^n},$$

all evaluated at (ξ_1,ξ_2).

If we write $x_1-a=h_1$, and $x_2-b=h_2$, and change to polar coordinates so that $h_1 = r\cos\alpha$ and $h_2 = r\sin\alpha$, then the first equation in (2.2.1) becomes

$$\Delta f = f(a+r\cos\alpha, b+r\sin\alpha)-f(a,b)=r\Big\{ f_1(a,b)\cos\alpha+f_2(a,b)\sin\alpha$$

$$+\frac{r}{2}\big[f_1(\xi_1,\xi_2)\cos^2\alpha+2f_{12}(\xi_1,\xi_2)\sin\alpha\cos\alpha+f_{22}(\xi_1,\xi_2)\sin^2\alpha\big]\Big\}. \quad (2.2.3)$$

As in the case of a function of one variable, the expression within the square brackets can be shown to be bounded and may be made arbitrarily small by choosing a suitable neighborhood around the point (a,b). The sign of the right-hand side of (2.2.3) is thus dominated by the first partial derivatives $f_1(a,b)$ and $f_2(a,b)$. If one of them, say $f_1(a,b)$, is equal to a positive number c then the entire right-hand side can be made positive by choosing, for instance, $\alpha=0$ after having made the third term less than c. In this case $\Delta f>0$ and $f(a,b)$ cannot be a maximum. It is not difficult to see that if $f(a,b)$ is to be a maximum (minimum), a necessary condition is that the first partial derivatives should both be equal to zero, for otherwise the sign of the right-hand side of (2.2.3) can be made to vary by choosing points in the neighborhood of (a,b) such that $\Delta f>0$ and others such that $\Delta f<0$ (by rotating the angle α).

A similar line of reasoning may be pursued in the case of a function $f(x_1,x_2,\ldots,x_n)$ of several variables, and we can in fact state as a theorem that a necessary condition for such a function to attain a maximum (minimum) is that each of $\partial f/\partial x_1, \partial f/\partial x_2,\ldots,\partial f/\partial x_n$ evaluated at $(a_1,a_2,\ldots,a_n)$ is zero.

2.3 Sufficient condition for a function of two variables

Suppose that the first partial derivatives at (a,b) are both zero. The second equation of (2.2.3) now reads

$$\Delta f = \frac{r^2}{2} \left(A \cos^2\alpha + 2B \sin\alpha \cos\alpha + C \sin^2\alpha + \frac{r}{3} D \right), \qquad (2.3.1)$$

where

$$A = f_{11}(a,b),$$
$$B = f_{12}(a,b),$$
$$C = f_{22}(a,b),$$

and D is a function which is bounded and may be made arbitrarily small. If $A \neq 0$, (2.3.1) may be rewritten as

$$\Delta f = \frac{r^2}{2} \left[\frac{(A \cos\alpha + B \sin\alpha)^2 + (AC - B^2) \sin^2\alpha}{A} + \frac{r}{3} D \right]. \qquad (2.3.2)$$

There are the following possible cases:

(a) $AC - B^2 > 0$. This implies that A and C are of the same sign. In this case, since the term $(r/3)D$ is dominated by the first fraction in the square brackets, the sign of Δf is the same as that of A. Thus, we have a minimum at the point (a,b) if $A > 0$ (hence $C > 0$ and $\Delta f > 0$) and a maximum if $A < 0$.

(b) $AC - B^2 < 0$. The sign of the fraction

$$\frac{(A \cos\alpha + B \sin\alpha)^2 + (AC - B^2) \sin^2\alpha}{A},$$

hence also the sign of Δf, will be the same as A for certain values of α (for example, $\alpha = 0$) and opposite to that of A for certain other values of α. For example, if α is chosen to be the angle for which $\tan\alpha = -A/B$, then $(A \cos\alpha + B \sin\alpha)^2$ vanishes. Since Δf changes sign for different points in the neighborhood of (a,b), we cannot have an extremum. Instead, the function has a saddle point at (a,b).

(c) $AC - B^2 = 0$. Now the numerator of the fraction is either zero or positive and we cannot say anything about the sign of Δf without investigating the function D, namely, the terms involving the higher derivatives.

In the above three cases it was assumed that $A \neq 0$. If $A = 0$, we may look at (2.3.1), from which we see that

$$2B \sin\alpha \cos\alpha + C \sin^2\alpha$$

may vanish for some values of α (such as $\alpha = 0$). Again we will find it necessary to investigate the function D. Unfortunately, the study of the higher derivatives is much too involved to be considered here.

In summary, then, for a function $f(x_1, x_2)$ of two variables, if the first partial derivatives evaluated at (a, b) are zero, we examine

$$E = f_{11}(a, b) \cdot f_{22}(a, b) - (f_{12}(a, b))^2$$

to determine whether there is a local extremum.

(1) If $E > 0$ and $f_{11}(a, b) > 0$ (hence $f_{22}(a, b)$ also > 0), we have a minimum.
(2) If $E > 0$ and $f_{11}(a, b) < 0$ (hence $f_{22}(a, b) < 0$) we have a maximum.
(3) If $E < 0$, and $f_{11}(a, b) \cdot f_{22}(a, b) \neq 0$, there is no maximum or minimum but a saddle point.
(4) If $E = 0$, or if $f_{11}(a, b) \cdot f_{22}(a, b) = 0$, we have to examine the higher derivatives or investigate the functional values at and near (a, b).

Example 1. Let $f(x, y) = x^2 + y^3 - 27y$. Then: $f_1 = \partial f / \partial x = 2x$; $f_2 = \partial f / \partial y = 3y^2 - 27$; $f_{11} = \partial^2 f / \partial x^2 = 2$; $f_{22} = \partial^2 f / \partial y^2 = 6y$; $f_{12} = \partial^2 f / \partial x \partial y = 0$.
Solving the equations involving the first-order derivatives gives the points $(0, 3)$ and $(0, -3)$.
At $(0, 3)$, $f_{11} \cdot f_{22} - f_{12}^2 = (2)(18) - 0 > 0$, and $f_{11}(0, 3) = 2 > 0$. Hence the point $(0, 3)$ gives a minimum.
At $(0, -3)$, $f_{11} \cdot f_{22} - f_{12}^2 = (2)(-18) - 0 < 0$. There is no maximum or minimum but a saddle point.

Example 2. Let $f = (x - y)^4 + (y - 2)^4$. Then: $f_1 = 4(x - y)^3$; $f_2 = 4(y - 2)^3 - 4(x - y)^3$; $f_{11} = 12(x - y)^2$; $f_{22} = 12(y - 2)^2 + 12(x - y)^2$; $f_{12} = -12(x - y)^2$.
At $x = y = 2$, $f_1 = f_2 = 0$, so that the point $(2, 2)$ may be an extremum. But $f_{11}(2, 2) = 0$, so we cannot use the sufficiency conditions but have to examine the value of the function near the point $(2, 2)$.
Let h and k be any arbitrarily small numbers and let us consider $f(x + h, y + k)$ at $x = y = 2$. $f(2 + h, 2 + k) = (h - k)^4 + k^4$, which is always positive and greater than $f(2, 2)$. Thus, $f(2, 2) = 0$ is a minimum.

Example 3. Let the following symbols be used for the sale of a certain commodity Q whose production requires labor (L) and capital (K):

$k =$ units of capital input;
$l =$ units of labor input;
$q = q(k, l) =$ output of the commodity Q (production function);
$p =$ unit sales price of output;
r_1 and $r_2 =$ unit prices of inputs of capital and labor.

Total profit π is equal to total revenue minus total cost, *i.e.*,

$$\pi = pq - (r_1 k + r_2 l).$$

Suppose that $p = 10$, $r_1 = 2$, $r_2 = 4$, and

$$q = 24k + 32l - k^2 - l^2.$$

Then

$$\pi = 10(24k + 32l - k^2 - l^2) - (2k + 4l);$$

$$\frac{\partial \pi}{\partial k} = 238 - 20k; \qquad \frac{\partial \pi}{\partial l} = 316 - 20l;$$

$$\frac{\partial^2 \pi}{\partial k^2} = \frac{\partial^2 \pi}{\partial l^2} = -20; \qquad \frac{\partial^2 \pi}{\partial k \partial l} = 0;$$

$$\frac{\partial^2 \pi}{\partial k^2} \cdot \frac{\partial^2 \pi}{\partial l^2} - \left(\frac{\partial^2 \pi}{\partial k \partial l}\right)^2 = 400 > 0;$$

$$\frac{\partial^2 \pi}{\partial k^2} < 0; \qquad \frac{\partial^2 \pi}{\partial l^2} < 0.$$

So π has a maximum at $k = 11.9$ and $l = 15.8$.

Example 4. Consider a monopolist who has two production plants and sells in a single market. His profit is the difference between his total revenue and his total production costs for both plants:

$$\pi = p(q_1 + q_2) - c_1(q_1) - c_2(q_2),$$

where q_1, q_2 are the quantities which he produces at the two plants, price p is a function of his output, i.e., $p = 200 - q_1 - q_2$, and $c_1(q_1) = 10q_1$ and $c_2(q_2) = q_2^2$ are his cost functions. Accordingly,

$$\pi = (200 - q_1 - q_2)(q_1 + q_2) - 10q_1 - q_2^2$$

$$= 190q_1 + 200q_2 - q_1^2 - 2q_2^2 - 2q_1q_2,$$

$$\frac{\partial \pi}{\partial q_1} = 190 - 2q_1 - 2q_2 = 0,$$

and

$$\frac{\partial \pi}{\partial q_2} = 200 - 4q_2 - 2q_1 = 0.$$

Solving these two equations together gives $q_1 = 90$ and $q_2 = 5$. Thus:

$$\frac{\partial^2 \pi}{\partial q_1^2} = -2; \qquad \frac{\partial^2 \pi}{\partial q_2^2} = -4; \qquad \frac{\partial^2 \pi}{\partial q_1 \partial q_2} = -2;$$

$$\frac{\partial^2 \pi}{\partial q_1^2} \cdot \frac{\partial^2 \pi}{\partial q_2^2} - \left(\frac{\partial^2 \pi}{\partial q_1 \partial q_2}\right)^2 = 4 > 0;$$

also,

$$\frac{\partial^2 \pi}{\partial q_1^2} < 0 \quad \text{and} \quad \frac{\partial^2 \pi}{\partial q_2^2} < 0.$$

We have a maximum. At $q_1 = 90$, $q_2 = 5$, $\pi = 9050$.

Example 5 (See Example 5, Section 1.5). To get the maximum likelihood estimators of the mean μ and variance σ^2 of the normal probability distribution

$$f(x) = \frac{1}{\sqrt{2\pi}\,\sigma}\exp\left(-\frac{1}{2\sigma^2}(x-\mu)^2\right), \qquad -\infty < \mu < \infty, \sigma > 0,$$

again we form the likelihood function $L(\mu,\sigma^2)$, where

$$L(\mu,\sigma^2) = \frac{1}{(2\pi)^{n/2}\sigma^n}\exp\left(-\frac{1}{2\sigma^2}\sum_{i=1}^{n}(x_i-\mu)^2\right),$$

and find the values of μ and σ^2 which maximize L or $\log L$.

$$\log L = -\frac{n}{2}\log 2\pi - n\log\sigma - \frac{1}{2\sigma^2}\sum_{i=1}^{n}(x_i-\mu)^2;$$

$$\frac{\partial \log L}{\partial \mu} = \frac{1}{\sigma^2}\sum_{i=1}^{n}(x_i-\mu) = \frac{1}{\sigma^2}(\Sigma x_i - n\mu) = 0; \qquad \mu = \frac{\Sigma x_i}{n} = \bar{x};$$

$$\frac{\partial \log L}{\partial \sigma} = -\frac{n}{\sigma} + \frac{\Sigma(x_i-\mu)^2}{\sigma^3} = 0; \qquad \sigma^2 = \frac{\Sigma(x_i-\mu)^2}{n}.$$

Solving the above two equations together gives

$$\mu = \bar{x} \quad \text{and} \quad \sigma^2 = \frac{1}{n}\sum_{i=1}^{n}(x_i-\bar{x})^2.$$

$$\frac{\partial^2 \log L}{\partial \mu^2} = -\frac{n}{\sigma^2} < 0 \quad \text{at}\left(\bar{x}, \frac{1}{n}\Sigma(x_i-\bar{x})^2\right).$$

Again, at this point,

$$\frac{\partial^2 \log L}{\partial \sigma^2} = \frac{n}{\sigma^2} - \frac{3\Sigma(x_i-\mu)^2}{\sigma^4} = \frac{1-3n^2}{\Sigma(x_i-\bar{x})^2} < 0$$

and

$$\frac{\partial^2 \log L}{\partial \sigma \partial \mu} = -\frac{2(\Sigma x_i - n\bar{x})}{\sigma^3} = 0,$$

so that

$$\left(\frac{\partial^2 \log L}{\partial \mu^2}\right)\left(\frac{\partial^2 \log L}{\partial \sigma^2}\right) - \left(\frac{\partial^2 \log L}{\partial \sigma \partial \mu}\right)^2 > 0,$$

and we have a maximum. Thus,

$$\hat{\mu} = \bar{x} \quad \text{and} \quad \hat{\sigma}^2 = \frac{1}{n}\sum_{i=1}^{n}(x_i-\bar{x})^2.$$

Example 6. The method of least squares is used frequently in statistics and applied problems. If we make n observations of the values of two variables

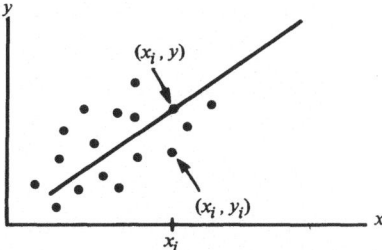

Figure 2.3.1.

x and y, we would get n points $(x_1,y_1),(x_2,y_2),\ldots,(x_n,y_n)$, such as those shown in Figure 2.3.1. If it appears that an approximate linear relationship exists between the two variables, then, to measure the degree of proportionality between x and y, we try to fit the set of points by passing a straight line

$$y = ax + b$$

through these points. The slope of the line is a and the y-intercept is b. The line which most closely fits these points should be the totality of points (x,y) such that the sum of the vertical distances from each observed point (x_i,y_i) to its corresponding point (x_i,y) on the line will be a minimum. One approach is to determine the parameters a and b so that

$$\sum_{i=1}^{n} |y - y_i| = \sum_{i=1}^{n} |ax_i + b - y_i|$$

is a minimum. Absolute values are often difficult to handle, and, instead, it is easier to determine a and b so that

$$S(a,b) = \sum_{i=1}^{n} (y - y_i)^2 = \sum_{i=1}^{n} (ax_i + b - y_i)^2$$

is a minimum. The reason we consider the sum of the absolute differences or the sum of the squares of the differences, instead of simply $\sum_{i=1}^{n}(y - y_i)$, is that in $\Sigma(y - y_i)$, some of the differences are positive while some are negative, depending on whether the points (x_i,y_i) lie above or below the line. Therefore a simple sum of the differences will not give a true picture of the amount of deviation of the line from the observed points.

$$\frac{\partial S}{\partial a} = 2 \sum_{i=1}^{n} (ax_i + b - y_i)x_i; \qquad \frac{\partial S}{\partial b} = 2 \sum_{i=1}^{n} (ax_i + b - y_i);$$

$$\frac{\partial^2 S}{\partial a^2} = 2 \sum_{i=1}^{n} x_i^2 > 0 \quad \text{and} \quad \frac{\partial^2 S}{\partial b^2} = 2 \sum_{i=1}^{n} 1 = 2n > 0;$$

$$\frac{\partial^2 S}{\partial a \partial b} = 2 \sum_{i=1}^{n} x_i; \qquad \left(\sum_{i=1}^{n} x_i^2\right)(n) - \left(\sum_{i=1}^{n} x_i\right)^2 = \frac{1}{2} \sum_{i=1}^{n} \sum_{j=1}^{n} (x_i - x_j)^2 > 0.$$

Hence S has a minimum for the pair of values of a and b which satisfy the equations

$$\Sigma(ax_i + b - y_i)x_i = 0$$

and

$$\Sigma(ax_i + b - y_i) = 0.$$

Again writing $\bar{x} = \Sigma x_i/n$ and $\bar{y} = \Sigma y_i/n$ as we have done in a previous example, the above two equations become

$$a\Sigma x_i^2 + bn\bar{x} - \Sigma x_i y_i = 0 \quad \text{and} \quad an\bar{x} + nb - n\bar{y} = 0,$$

so that

$$a = \frac{\Sigma x_i y_i - n\bar{x}\bar{y}}{\Sigma x_i^2 - n(\bar{x})^2} \quad \text{and} \quad b = \bar{y} - \frac{\Sigma x_i y_i - n\bar{x}\bar{y}}{\Sigma x_i^2 - n(\bar{x})^2}\bar{x}.$$

In actual practice, we frequently shift the graph so that the origin falls on the point $(\bar{x}, \bar{y})$. Then, instead of the line $y = ax + b$, we consider the line

$$y = a(x - \bar{x}) + b.$$

This does not change the slope of the line, of course, but the vertical or y-intercept is now on the line $x = \bar{x}$, that is,

$$S(a,b) = \Sigma\left[a(x_i - \bar{x}) + b - y_i\right]^2,$$

and exactly the same calculations yield

$$b = \bar{y} \quad \text{and} \quad a = \frac{\Sigma x_i y_i - n\bar{x}\bar{y}}{\Sigma x_i^2 - n(\bar{x})^2}.$$

The least squares line through the observed points is of the form

$$y - \bar{y} = \frac{\Sigma x_i y_i - n\bar{x}\bar{y}}{\Sigma x_i^2 - n(\bar{x})^2}(x - \bar{x}).$$

This line is called the *regression line of y on x*.

In the above formulation we considered the vertical deviations of the points on the line from the observed points. We assumed that the variable x could be measured quite accurately whereas y was subject to random errors. If, instead, we treat x as a function of y and consider the horizontal deviations, we would be interested in a line of the form

$$x = a'(y - \bar{y}) + b'$$

and we would get a corresponding regression line of x on y.

Another result that is of interest is that, since $n\bar{x} = \Sigma x_i$ and $n\bar{y} = \Sigma y_i$,

$$a = \frac{\Sigma(x_i - \bar{x})(y_i - \bar{y})}{\Sigma(x_i - \bar{x})^2}.$$

The expressions $s_x^2 = (1/n)(x_i - \bar{x})^2$ and $s_y^2 = (1/n)(y_i - \bar{y})^2$ are called the *sample variances of* x *and* y respectively, and the sample correlation coefficient r is

$$r = \frac{\Sigma(x_i - \bar{x})(y_i - \bar{y})}{ns_x s_y} = \frac{\Sigma(x_i - \bar{x})(y_i - \bar{y})}{\sqrt{\Sigma(x_i - \bar{x})^2}\sqrt{\Sigma(y_i - \bar{y})^2}}.$$

Therefore

$$a = \frac{rs_y}{s_x}$$

and the regression line of y on x is

$$y - \bar{y} = \frac{rs_y}{s_x}(x - \bar{x}), \quad i.e., \quad \frac{y - \bar{y}}{s_y} = r\frac{x - \bar{x}}{s_x}.$$

Example 7. Suppose we wish to find the equation of the least squares line through the points $(1,1)$, $(1,3)$, $(2,6)$, $(2.5,4)$, $(3,7.5)$. $\bar{x} = 1.9$, $\bar{y} = 4.3$, $\Sigma x_i y_i = 48.5$, $\Sigma x_i^2 = 21.25$, $n\bar{x}\bar{y} = 40.85$, and $n\bar{x}^2 = 18.05$. (See Figure 2.3.2.)

The equation of the line is

$$y - 4.3 = \frac{48.5 - 40.85}{21.25 - 18.05}(x - 1.9),$$

i.e.,

$$y = 2.39x - 0.24.$$

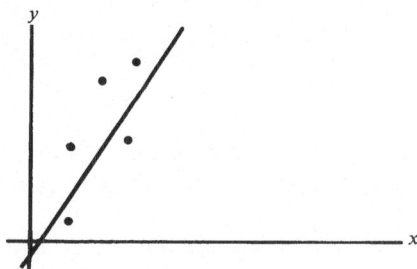

Figure 2.3.2.

2.4 Sufficient condition for a function $f(x_1, x_2, \ldots, x_n)$ of n variables

We shall use the same notations as previously, that is,

$$f_i(a_1, a_2, \ldots, a_n) \quad \text{stands for} \quad \frac{\partial f(x_1, x_2, \ldots, x_n)}{\partial x_i}$$

and

$$f_{ij}(a_1, a_2, \ldots, a_n) \quad \text{stands for} \quad \frac{\partial^2 f(x_1, x_2, \ldots, x_n)}{\partial x_i \partial x_j},$$

each evaluated at the point $(a_1, a_2, \ldots, a_n)$. Also, $h_i = x_i - a_i$, $\xi_i = a_i + \theta h_i$.

Let us first consider a function $f(x_1, x_2, x_3)$ of three variables which has continuous first- and second-order partial derivatives at and in a neighborhood of (a_1, a_2, a_3). Again, by Taylor's theorem,

$$\Delta f = f(x_1, x_2, x_3) - f(a_1, a_2, a_3)$$

$$= \sum_{i=1}^{3} f_i(a_1, a_2, a_3) h_i + \frac{1}{2} \sum_{i=1}^{3} \sum_{j=1}^{3} f_{ij}(\xi_1, \xi_2, \xi_3) h_i h_j. \tag{2.4.1}$$

Suppose the first partial derivatives at (a_1, a_2, a_3) are zero; then

$$\Delta f = \frac{1}{2} \sum\sum f_{ij} h_i h_j$$

$$= \frac{1}{2}\left(f_{11} h_1^2 + 2 f_{12} h_1 h_2 + 2 f_{13} h_1 h_3 + f_{22} h_2^2 + 2 f_{23} h_2 h_3 + f_{33} h_3^2 \right). \tag{2.4.2}$$

The expression within parentheses in (2.4.2) is a *quadratic form in the three variables* h_1, h_2, h_3.

In general, by a quadratic form in any three variables, say, u_1, u_2, and u_3, we mean

$$F(u_1, u_2, u_3) = \sum_{i=1}^{3} \sum_{j=1}^{3} b_{ij} u_i u_j$$

$$= b_{11} u_1^2 + 2 b_{12} u_1 u_2 + 2 b_{13} u_1 u_3 + b_{22} u_2^2 + 2 b_{23} u_2 u_3 + b_{33} u_3^2,$$

with $b_{ij} = b_{ji}$.

F is *positive definite* if and only if $F(u_1, u_2, u_3) > 0$, except when $u_1 = u_2 = u_3 = 0$. F is *negative definite* if and only if $F < 0$, except when $u_1 = u_2 = u_3 = 0$. Clearly, $F(0, 0, 0) = 0$.

It is obvious that if we wish $f(a_1, a_2, a_3)$ to be a maximum, then Δf must be negative. That is, we want the quadratic form (2.4.2) to be negative definite.

We rewrite that quadratic form as follows:

$$2\Delta f = f_{11}\left(h_1 + \frac{f_{12}}{f_{11}} h_2 + \frac{f_{13}}{f_{11}} h_3 \right)^2$$

$$+ \frac{f_{11} f_{22} - f_{12}^2}{f_{11}} \left(h_2 - \frac{f_{13} f_{12} - f_{11} f_{23}}{f_{11} f_{22} - f_{12}^2} h_3 \right)^2 \tag{2.4.3}$$

$$+ \frac{f_{11} f_{22} f_{33} + 2 f_{23} f_{13} f_{12} - f_{11} f_{23}^2 - f_{22} f_{13}^2 - f_{33} f_{12}^2}{f_{11} f_{22} - f_{12}^2} h_3^2,$$

which, by using determinants, can be written

$$2\Delta f = f_{11}\left(h_1 + \frac{f_{12}}{f_{11}}h_2 + \frac{f_{13}}{f_{11}}h_3\right)^2$$

$$+ \frac{\begin{vmatrix} f_{11} & f_{12} \\ f_{12} & f_{22} \end{vmatrix}}{f_{11}}\left[h_2 - \frac{f_{13}f_{12} - f_{11}f_{23}}{\begin{vmatrix} f_{11} & f_{12} \\ f_{12} & f_{22} \end{vmatrix}}h_3\right]^2 + \frac{\begin{vmatrix} f_{11} & f_{12} & f_{13} \\ f_{12} & f_{22} & f_{23} \\ f_{13} & f_{23} & f_{33} \end{vmatrix}}{\begin{vmatrix} f_{11} & f_{12} \\ f_{12} & f_{22} \end{vmatrix}}h_3^2. \qquad (2.4.4)$$

Expression (2.4.4) is essentially the sum of three squares each multiplied by a coefficient. If the coefficients are all negative, then surely Δf will be negative. This occurs if

$$f_{11} < 0, \quad \begin{vmatrix} f_{11} & f_{12} \\ f_{12} & f_{22} \end{vmatrix} > 0 \quad \text{and} \quad D_3 = \begin{vmatrix} f_{11} & f_{12} & f_{13} \\ f_{12} & f_{22} & f_{23} \\ f_{13} & f_{23} & f_{33} \end{vmatrix} < 0. \quad (2.4.5)$$

Similarly, Δf must be positive for $f(a_1, a_2, a_3)$ to be a minimum. This occurs if the coefficients in (2.4.4) are all positive, that is, if

$$f_{11} > 0, \quad \begin{vmatrix} f_{11} & f_{12} \\ f_{12} & f_{22} \end{vmatrix} > 0, \quad \text{and} \quad D_3 = \begin{vmatrix} f_{11} & f_{12} & f_{13} \\ f_{12} & f_{22} & f_{23} \\ f_{13} & f_{23} & f_{33} \end{vmatrix} > 0. \quad (2.4.6)$$

Actually, the above three conditions of (2.4.5) or (2.4.6) can be shown to be not only sufficient, but also necessary for a function $f(x_1, x_2, x_3)$ to attain a strict local maximum (or minimum). The interested reader is referred to books on advanced calculus and quadratic forms for the proof.

The expressions (2.4.2), (2.4.3), and (2.4.4) can be rearranged in various ways. For example,

$$2\Delta f = f_{33}h_3^2 + f_{23}h_2h_3 + f_{13}h_1h_3 + f_{23}h_2h_3 + f_{22}h_2^2 + f_{12}h_1h_2$$
$$+ f_{13}h_1h_3 + f_{12}h_1h_2 + f_{11}h_1^2 \qquad (2.4.7)$$

$$= f_{33}\left(h_3 + \frac{f_{23}}{f_{33}}h_2 + \frac{f_{13}}{f_{33}}h_1\right)^2 + \frac{\begin{vmatrix} f_{33} & f_{23} \\ f_{23} & f_{22} \end{vmatrix}}{f_{33}}\left(h_2 - \frac{f_{13}f_{23} - f_{33}f_{12}}{f_{33}f_{22} - f_{23}^2}h_1\right)^2$$

$$+ \frac{\begin{vmatrix} f_{33} & f_{23} & f_{13} \\ f_{23} & f_{22} & f_{12} \\ f_{13} & f_{12} & f_{11} \end{vmatrix}}{\begin{vmatrix} f_{33} & f_{23} \\ f_{23} & f_{22} \end{vmatrix}}h_1^2. \qquad (2.4.8)$$

By exactly the same reasoning as above, the conditions given by (2.4.5) and (2.4.6) can be made to read:

$$f_{33}<0, \quad \begin{vmatrix} f_{33} & f_{23} \\ f_{23} & f_{22} \end{vmatrix} >0, \quad \text{and} \quad D_3 = \begin{vmatrix} f_{33} & f_{23} & f_{13} \\ f_{23} & f_{22} & f_{12} \\ f_{13} & f_{12} & f_{11} \end{vmatrix} <0; \quad (2.4.9)$$

and

$$f_{33}>0, \quad \begin{vmatrix} f_{33} & f_{23} \\ f_{23} & f_{22} \end{vmatrix} >0, \quad \text{and} \quad D_3>0. \quad (2.4.10)$$

The determinant D_3 of (2.4.5), (2.4.6), (2.4.9), and (2.4.10) is of course the same by interchanging the first and third rows and corresponding columns. It is called the *determinant* or *discriminant of the quadratic form*.

In general, consider a determinant Δ_n of *order n*, where

$$\Delta_n = \begin{vmatrix} a_{11} & a_{12} & \cdots & a_{1n} \\ \vdots & & & \\ a_{n1} & a_{n2} & \cdots & a_{nn} \end{vmatrix}.$$

If we delete the ith row and the ith column, we obtain a determinant of order $n-1$. We call this second determinant a *principal minor* of Δ_n.

We have thus arrived at the necessary and sufficient conditions for a function $f(x_1, x_2, x_3)$ of three variables to attain a strict local extremum. We summarize them: If

(1) $f(x_1, x_2, x_3)$ has continuous first- and second-order partial derivatives,
(2) $f_i(a_i, a_2, a_3)=0$, $i=1,2,3$,
(3) the determinant D_3 of (2.4.9) is negative, all its principal minors of order 2 are positive and all its principal minors of order 1 are negative,

then $f(a_1, a_2, a_3)$ is a local maximum. If, in addition to conditions (1) and (2), condition (3) is changed to read "D_3 and its principal minors are positive," then $f(a_1, a_2, a_3)$ is a local minimum.

In the formulation which led to this result, we used $f_{ij}(\xi_1, \xi_2, \xi_3)$. Since the second-order partial derivatives are continuous, the statements about the signs of the determinants also hold at (a_1, a_2, a_3) (see Footnote 1). Therefore f_{ij} in D_3 stands for $f_{ij}(a_1, a_2, a_3)$.

The above reasoning can be easily applied to a function $f(x_1, x_2, \ldots, x_n)$ of n variables. We merely state the results:

If the function has continuous third-order partial derivatives at and in a neighborhood of $(a_1, a_2, \ldots, a_n)$ and $f_i(a_1, a_2, \ldots, a_n)=0$ for $i=1,2,\ldots,n$, then

---

[1]Continuity implies that if a continuous function is positive (or negative) at a point, then it will have the same sign in a small neighborhood of that point.

Table 2.4.1

Maximum	$D_n < 0$ for n odd > 0 for n even	Principal minors of odd order < 0	Principal minors of even order > 0
Minimum	D_n and all principal minors < 0		

we look at the determinant

$$D_n = \begin{vmatrix} f_{11} & f_{12} \cdots & f_{1n} \\ f_{12} & \cdots & f_{2n} \\ \vdots & & \\ f_{1n} & \cdots & f_{nn} \end{vmatrix}$$

of the quadratic form, where f_{ij} stand for $\partial^2 f / \partial x_i \partial x_j$ evaluated at the point $(a_1, a_2, \ldots, a_n)$. $f(a_1, a_2, \ldots, a_n)$ is a minimum if D_n and all its principal minors are positive. $f(a_1, a_2, \ldots, a_n)$ is a maximum if all the principal minors of order 1 of D_n are negative, of order 2 are positive, and principal minors of increasing order alternate in sign, finally with D_n negative for n odd and positive for n even. This is summarized in Table 2.4.1.

An alternate approach is to look at the characteristic equation of D_n (see Footnote 2),

$$\begin{vmatrix} f_{11} - r & f_{12} \cdots & f_{1n} \\ f_{12} & f_{22} - r & f_{2n} \\ \vdots & & \\ f_{1n} & \cdots & f_{nn} - r \end{vmatrix} = 0,$$

and the roots of the above equation, which, when expanded, is a polynomial equation of degree n in r. We call the greatest root r_g and the smallest root r_s.

(1) If $r_s > 0$, $f(a_1, a_2, \ldots, a_n)$ is a minimum.
(2) If $r_g < 0$, $f(a_1, a_2, \ldots, a_n)$ is a maximum.
(3) If $r_s < 0$ and $r_g > 0$, there is no extremum.
(4) If $r_s \cdot r_g = 0$, we have to examine the function or its higher derivatives.

The reader should see that the sufficient conditions arrived at in Section 2.3 for a function of two variables is really just a special case of the general result, with $n = 2$.

Example 1. Given three isolated markets supplied by a single monopolist, let the three corresponding demand functions be

$$p_1 = 12 - q_1, \qquad p_2 = 18 - 2q_2, \qquad p_3 = 20 - 3q_2,$$

[2]See Appendix I for a discussion of the characteristic roots of a matrix.

and let his cost function be $C = 3 + 2(q_1 + q_2 + q_3)$.

The total profit from the three markets will be

$$\pi = p_1 q_1 + p_2 q_2 + p_3 q_3 - C$$
$$= (12 - q_1)q_1 + (18 - 2q_2)q_2 + (20 - 3q_3)q_3 - 3 - 2(q_1 + q_2 + q_3)$$
$$= 10q_1 + 16q_2 + 18q_3 - q_1^2 - 2q_2^2 - 3q_3^2 - 3.$$

Thus,

$$\frac{\partial \pi}{\partial q_1} = 10 - 2q_1 = 0 \qquad q_1 = 5,$$

$$\frac{\partial \pi}{\partial q_2} = 16 - 4q_2 = 0 \qquad q_2 = 4,$$

$$\frac{\partial \pi}{\partial q_3} = 18 - 6q_3 = 0 \qquad q_3 = 3,$$

$$\frac{\partial^2 \pi}{\partial q_1^2} = -2, \qquad \frac{\partial^2 \pi}{\partial q_2^2} = -4, \qquad \frac{\partial^2 \pi}{\partial q_3^2} = -6,$$

$$\frac{\partial^2 \pi}{\partial q_1 \partial q_2} = 0, \qquad \frac{\partial^2 \pi}{\partial q_1 \partial q_3} = 0, \qquad \frac{\partial^2 \pi}{\partial q_2 \partial q_3} = 0,$$

$$\begin{vmatrix} -2 & 0 \\ 0 & -4 \end{vmatrix} = 8 > 0, \qquad \begin{vmatrix} -2 & 0 & 0 \\ 0 & -4 & 0 \\ 0 & 0 & -6 \end{vmatrix} = -48 < 0.$$

$\pi = 81$ is a maximum at $q_1 = 5$, $q_2 = 4$ and $q_3 = 3$.

Example 2. Let $f(x_1, x_2, x_3, x_4) = 24x_1 + 32x_2 + 48x_3 + 72x_4 - (x_1^2 + x_2^2 + 2x_3^2 + 3x_4^2)$. Then

$$\frac{\partial f}{\partial x_1} = 24 - 2x_1, \qquad \frac{\partial f}{\partial x_2} = 32 - 2x_2,$$

$$\frac{\partial f}{\partial x_3} = 48 - 4x_3, \qquad \frac{\partial f}{\partial x_4} = 72 - 6x_4.$$

Solving $\partial f / \partial x_i = 0$, we get $x_1 = 12$, $x_2 = 16$, $x_3 = 12$, $x_4 = 12$. $\partial^2 f / \partial x_1^2 = -2$, $\partial^2 f / \partial x_2^2 = -2$, $\partial^2 f / \partial x_3^2 = -4$, $\partial^2 f / \partial x_4^2 = -6$, and $\partial^2 f / \partial x_i \partial x_j = 0$ for $i \neq j$; thus,

$$D_4 = \begin{vmatrix} -2 & 0 & 0 & 0 \\ 0 & -2 & 0 & 0 \\ 0 & 0 & -4 & 0 \\ 0 & 0 & 0 & -6 \end{vmatrix} > 0.$$

We see that all principal minors of order 1 and 3 are negative, while all principal minors of order 2 are positive. $f(12, 16, 12, 12) = 1120$ is a maximum.

Exercises

2.1 Let $p_1 = 52$ and $p_2 = 44$ be the prices of two products, and let q_1 and q_2 be the quantities of these products, respectively, under pure competition. The firm's revenue function, then, is $R = 52q_1 + 44q_2$. Let the cost function be $C = q_1^2 + q_1 q_2 + q_2^2$. Profit $\pi = R - C$. Find the quantities q_1 and q_2 that this firm should produce in order to maximize profit.

2.2 Let the demand functions of a two-product monopolist be

$$p_1 = 256 - 3q_1 - q_2,$$
$$p_2 = 222 + q_1 - 5q_2,$$

where p_1, q_1, p_2, q_2 are the prices and quantities of the two products respectively. Let the cost function be

$$C = q_1^2 + q_1 q_2 + q_2^2.$$

Find the quantities q_1 and q_2 that should be produced to maximize profit.

2.3 Given the function $f(x_1, x_2, x_3) = x_1^2 + x_1 x_2 + 2x_2^2 + x_1 x_3 + 4x_3^2 + 5$. Find its extreme value(s), if any.

2.4 The following example[3] shows how the maximization of a function of two variables may be transformed into the maximization of a function of one variable. Let a consumer's satisfaction be a function of leisure L and his purchasing power (income) y. That is, his utility function is $u = g(L, y)$. The rate of substitution of income for leisure is defined to be

$$-\frac{dy}{dL} = \frac{\dfrac{\partial g}{\partial L}}{\dfrac{\partial g}{\partial y}}.$$

Let T = total amount of time available to him, W = amount of work performed by him, and r = wage rate. Then $L = T - W$, $y = rW$, and $u = g(T - W, rW)$. Assume that T and r are given. (a) Find the necessary condition for maximizing u. (b) Suppose $u = Ly$. How much work should he perform to achieve maximum utility?

2.5 Let the utility of a country for the consumption of two goods X and Y be $u = u(x, y)$, where x and y are, respectively, the amounts of X and Y consumed. If x_0 and y_0 represent domestic outputs of X and Y, we define the *production transformation function* $y_0 = g(x_0)$. The country can also obtain Y by exporting e units of X and importing p units of Y for each unit of X exported. Thus $x = x_0 - e$ and $y = y_0 + pe$. (a) Derive an expression for u by incorporating the above information. (b) Find the necessary condition for u to be maximized. (c) Show that at equilibrium

$$\frac{\dfrac{\partial u}{\partial x}}{\dfrac{\partial u}{\partial y}} = \frac{dg(x_0)}{dx_0} = -p.$$

[3]See Henderson and Quandt (1971), pp. 29–31.

2.6 Find the maximum likelihood estimators of μ and σ^2 of a normal distribution if six observed values of the variable are: -2.3; -2; -1; 1.8; 2.5; and 4.

2.7 Find the equation of the least squares line through the points $(1,1)$, $(2,0.8)$, $(3,1)$, $(4,0.2)$, $(5,0.7)$, $(6,1.2)$, $(7,1.4)$, $(8,0.9)$.

3 Functions of Two or More Variables (with Constraint)

3.1 Preliminary

In the problem of finding an extreme value of a function $f(x_1, x_2)$ of two variables, it may happen that the variables x_1 and x_2 are not independent but are connected by one or more relations called *constraints*. In theory, we can solve for one of the two variables in terms of the other, and proceed as in Chapter 1. In actual practice, however, we are held back by two considerations. First, it may be tedious to substitute one variable for another. Second, and more important, we have to be careful in choosing which one of x_1 and x_2 to be the independent variable. The following example illustrates the possible pitfalls when we are not careful.

Example 1. We wish to find the shortest distance from the point $(1,0)$ to the parabola $y^2 = 8x$. The problem here is to minimize

$$f(x,y) = (x-1)^2 + (y-0)^2$$

subject to the constraint

$$g(x,y) = y^2 - 8x = 0.$$

If we solve for y in $g(x,y)$ and substitute that in $f(x,y)$, we get $u(x) = (x-1)^2 + 8x$, which is to be minimized. Now, $du/dx = 2x + 6 = 0$ yields $x = -3$, which is absurd since there is no point on the parabola with a negative abscissa.

On the other hand, if we solve for x in terms of y and substitute that in $f(x,y)$, we get

$$v(y) = \left(\frac{y^2}{8} - 1\right)^2 + y^2,$$

$$\frac{dv}{dy} = \frac{y^3 + 24y}{16} = 0, \qquad y = 0,$$

$$\frac{d^2v}{dy^2} = \frac{3y^2 + 24}{16} > 0 \quad \text{at } y = 0.$$

Thus the shortest distance is to the point $(0,0)$ on the parabola.

The explanation to the difficulty demonstrated in the above example lies in the differentiation of implicit functions. In $g(x,y)=0$, if we take y as a function of x, we have

$$g_1 + g_2 \frac{dy}{dx} = 0, \quad i.e., \quad \frac{dy}{dx} = -\frac{g_1}{g_2}, \quad \text{provided } g_2 \neq 0,$$

where $g_1 = \partial g / \partial x$ and $g_2 = \partial g / \partial y$.

In the above example, $g_2 = 2y = 0$ at $y = 0$. Hence we have to take x as a function of y, since $g_1 = -8 \neq 0$. We then have

$$\frac{dv}{dy} = f_1 \frac{dx}{dy} + f_2 = -f_1 \frac{g_2}{g_1} + f_2.$$

3.2 Necessary condition

Let us consider the problem of finding the extreme value of a function $f(x_1, x_2)$ where the two variables x_1 and x_2 are related by the constraint

$$g(x_1, x_2) = 0,$$

and the partial derivatives g_1 and g_2 are not both zero. Graphically, the constraint may be represented by a curve $g=0$ in the $x_1 x_2$-plane as shown in Figure 3.2.1. It will intersect a family of curves $f(x_1, x_2) = c$. We wish to find that curve for which the value c_e is a maximum or minimum. As we trace the curve $g=0$, we cross the family of curves with the values of c increasing (or decreasing) monotonically. An extreme value occurs when the sense of c changes, and this occurs only when the two curves $f(x_1, x_2) = c_e$ and $g(x_1, x_2) = 0$ touch at a point (a_1, a_2). At that point of tangency, the normal vectors of the curves $g=0$ and $f=c_e$ are oriented in the same direction. The components of a normal vector are proportional to the partial derivatives of the function. Therefore, at the point (a_1, a_2), we have the proportional relation

$$f_1 : f_2 = g_1 : g_2, \tag{3.2.1}$$

where f_i and g_i stand for partial derivatives.

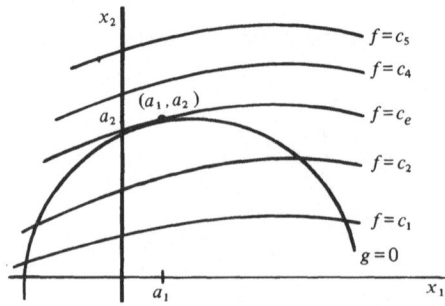

Figure 3.2.1.

Let λ be the constant of proportionality. Then we have

$$f_1 + \lambda g_1 = 0,$$
$$f_2 + \lambda g_2 = 0. \tag{3.2.2}$$

This constant λ is called the *Lagrange multiplier*. The two equations in (3.2.2), together with the equation

$$g(x_1, x_2) = 0,$$

constitute the necessary condition for $f(x_1, x_2)$ to attain an extremum. Moreover, these three equations serve to determine the three unknowns a_1, a_2, and λ.

The above necessary condition may be alternately formulated as follows: Form the function

$$L(x_1, x_2, \lambda) = f(x_1, x_2) + \lambda g(x_1, x_2),$$

where x_1, x_2, and the undetermined multiplier λ are treated as independent variables with no constraints. We then apply the necessary condition for the extremum of the function L as in Chapter 2 and set

$$\frac{\partial L}{\partial x_1} = f_1 + \lambda g_1 = 0,$$

$$\frac{\partial L}{\partial x_2} = f_2 + \lambda g_2 = 0, \tag{3.2.3}$$

and

$$\frac{\partial L}{\partial \lambda} = g(x_1, x_2) = 0.$$

This function L is called the *Lagrangian*.

The reasoning used for the case of a function of two variables with one constraint can be easily extended to the case of a function $f(x_1, x_2, \ldots, x_n)$ of n variables subject to m ($< n$) constraints

$$g^1(x_1, x_2, \ldots, x_n) = 0,$$
$$\vdots$$
$$g^m(x_1, x_2, \ldots, x_n) = 0,$$

where the partial derivatives $g_i^j = \partial g^j / \partial x_i$ ($i = 1, \ldots, n$; $j = 1, \ldots, m$) are not all zero. We also assume that the matrix of the partial derivatives has *row rank* equal to m; that is, the vectors $\nabla g^j = (\partial g^j / \partial x_1, \partial g^j / \partial x_2, \ldots, \partial g^j / \partial x_n)^{\mathrm{T}}$ are linearly independent.

We introduce m undetermined constants $\lambda_1, \ldots, \lambda_m$ and form the Lagrangian

$$L(x_1, \ldots, x_n, \lambda_1, \ldots, \lambda_m) = f(x_1, \ldots, x_n) + \sum_{j=1}^m \lambda_j g^j.$$

In order for f to attain an extremum, the n equations

$$\frac{\partial L}{\partial x_i} = f_i + \Sigma \lambda_j g_i^j = 0, \qquad i = 1, \ldots, n; j = 1, \ldots, m \tag{3.2.4}$$

plus the m equations

$$g^j = 0$$

must be satisfied.

This system of $m+n$ equations enables us to solve for the $m+n$ unknowns $x_1, \ldots, x_n, \lambda_1, \ldots, \lambda_m$.

We should remark that we have only presented a graphical, intuitive explanation of the Lagrange method of introducing the multipliers λ_j to establish the necessary condition for local extremum of the function f. For a more rigorous and analytic proof, we will have to solve for one of the two variables, say x_2, as a continuously differentiable function of x_1 in $g(x_1, x_2)$, substitute the expression $x_2 = h(x_1)$ in f, and proceed from there. We shall not undertake this here.[1]

3.3 Sufficient condition

We now consider the sufficient condition for a local extremum of the function $f(x_1, x_2)$ subject to the constraint $g(x_1, x_2) = 0$ in which the partial derivatives of f and g of all orders are continuous and g_1 and g_2 do not both vanish. Suppose the necessary conditions given by (3.2.3) are satisfied, and suppose, for convenience of discussion, that $g_2 \neq 0$. Then, $x_2 = h(x_1)$, and $f(x_1, h(x_1))$ can be regarded as a function $F(x_1)$ of x_1, so that

$$h' = \frac{dh(x_1)}{dx_1} = \frac{-g_1}{g_2}, \quad i.e., \quad g_1 + g_2 h' = 0,$$

which yields

$$\frac{d(g_1 g_2 h')}{dx_1} = g_{11} + 2g_{12}h' + g_{22}(h')^2 + g_2 h'' = 0.$$

In other words,

$$h'' = \frac{d^2 h(x_1)}{dx_1^2} = \frac{-1}{g_2^3} \left[g_{11}(g_2)^2 - 2g_{12}g_1 g_2 + g_{22}(g_1)^2 \right],$$

where $g_i = \partial g(x_1, x_2)/\partial x_i$ and $g_{ij} = \partial^2 g(x_1, x_2)/\partial x_i \partial x_j$. Then

$$\frac{dF}{dx_1} = f_1 + f_2 h', \tag{3.3.1}$$

and

$$\frac{d^2 F}{dx_1^2} = f_{11} + 2f_{12}h' + f_{22}(h')^2 + f_2 h''$$

$$= \frac{1}{(g_2)^2} \left[A_{11}(g_2)^2 - 2A_{12}g_1 g_2 + A_{22}(g_1)^2 \right],$$

[1]The interested reader is referred to Thomas (1972), Section 15.11, Kaplan (1973), Sections 2.19 and 2.20, or Aoki (1971), pp. 167–72.

where $A_{ij} = f_{ij} + \lambda g_{ij}$ of (3.2.3). By Taylor's formula,

$$\Delta F = F(x_1) - F(a_1) = \frac{1}{2}\left[\frac{d^2F}{dx_1^2}\bigg|_{x_1=a_1} + \delta\right](x_1 - a_1)^2 \qquad (3.3.2)$$

(hence $x_2 = a_2$) and δ can be made as small as desired in a neighborhood of (a_1, a_2). Therefore $f(a_1, a_2)$, which is $F(a_1)$, is a local maximum if $d^2F/dx_1^2 < 0$, and a minimum if $d^2F/dx_1^2 > 0$, at (a_1, a_2).

Exactly as in Section 2.3, the expression

$$E = A_{11}(g_2)^2 - 2A_{12}g_1 g_2 + A_{22}(g_1)^2,$$

which is a quadratic form in g_1 and g_2, can be rewritten as

$$E = \frac{(A_{11}g_2 - A_{12}g_1)^2 + (A_{11}A_{22} - A_{12}^2)(g_1)^2}{A_{11}}. \qquad (3.3.3)$$

If

$$A_{11} < 0 \quad \text{and} \quad \begin{vmatrix} A_{11} & A_{12} \\ A_{12} & A_{22} \end{vmatrix} = \begin{vmatrix} f_{11} + \lambda g_{11} & f_{12} + \lambda g_{12} \\ f_{12} + \lambda g_{12} & f_{22} + \lambda g_{22} \end{vmatrix} > 0,$$

we have $E < 0$, hence $d^2F/dx_1^2 < 0$ and $\Delta F < 0$, and $f(a_1, a_2)$ is a maximum. On the other hand, if both

$$A_{11} \quad \text{and} \quad \begin{vmatrix} A_{11} & A_{12} \\ A_{12} & A_{22} \end{vmatrix}$$

are positive, then $E > 0$, hence $\Delta F > 0$, and $f(a_1, a_2)$ is a minimum.

These conditions are sufficient, but not necessary, as we may have a maximum (minimum) with the determinant or A_{11} equal to zero.

When we move to a consideration of a function $f(x_1, x_2, x_3)$ of three variables, in which the conditions given by (3.2.4) are satisfied and which is subject to the constraint $g(x_1, x_2, x_3) = 0$, we will again be led to the second derivative.

The Taylor series expansion involves a rather complicated expression, and generally one uses a parameter, say t, so that the variables x_i can be thought of as functions of t. Then $f(x_1, x_2, x_3)$ and $g(x_1, x_2, x_3)$ both become functions of t, say $\alpha(t)$ and $\beta(t)$. Since $d^2\beta(t)/dt^2 = 0$ and the conditions given by (3.2.4) are satisfied, we would get

$$\frac{d^2\alpha(t)}{dt^2} = \sum_{i=1}^{3}(f_{ii} + \lambda g_{ii})\left(\frac{dx_i}{dt}\right)^2 + \sum_{i=1}^{3}\sum_{j=1}^{3}(f_{ij} + \lambda g_{ij})\frac{dx_i}{dt}\frac{dx_j}{dt} \qquad (3.3.4)$$

and we examine the sign of $d^2\alpha/dt^2$. If $d^2\alpha/dt^2 < 0$ we have a maximum, and if $d^2\alpha/dt^2 > 0$ we have a minimum.

This problem, as in Section 2.4, again involves the study of conditions under which a quadratic form

$$F(u_1, u_2, u_3) = \sum_i \sum_j a_{ij} u_i u_j \qquad (3.3.5)$$

is positive or negative definite. Here, when we consider a constraint, we also require that the quadratic form be positive or negative definite along the curve

$$p_1 u_1 + p_2 u_2 + p_3 u_3 = 0.$$

From algebra,[2] the conditions are: If

$$\begin{vmatrix} a_{11} & a_{12} & a_{13} & p_1 \\ a_{12} & a_{22} & a_{23} & p_2 \\ a_{13} & a_{23} & a_{33} & p_3 \\ p_1 & p_2 & p_3 & 0 \end{vmatrix} < 0 \tag{3.3.6}$$

and

$$\begin{vmatrix} a_{11} & a_{12} & p_1 \\ a_{12} & a_{22} & p_2 \\ p_1 & p_2 & 0 \end{vmatrix} < 0,$$

then $F(u_1, u_2, u_3)$ is positive. Converted to our problem, the conditions are: If

$$\begin{vmatrix} f_{11}+\lambda g_{11} & f_{12}+\lambda g_{12} & f_{13}+\lambda g_{13} & g_1 \\ f_{12}+\lambda g_{12} & f_{22}+\lambda g_{22} & f_{23}+\lambda g_{23} & g_2 \\ f_{13}+\lambda g_{13} & f_{23}+\lambda g_{23} & f_{33}+\lambda g_{33} & g_3 \\ g_1 & g_2 & g_3 & 0 \end{vmatrix} < 0 \tag{3.3.7}$$

and its border-preserving principal first minors (those of order 3 and formed by not deleting the fourth row or column) such as

$$\begin{vmatrix} f_{11}+\lambda g_{11} & f_{12}+\lambda g_{12} & g_1 \\ f_{12}+\lambda g_{12} & f_{22}+\lambda g_{22} & g_2 \\ g_1 & g_2 & 0 \end{vmatrix} < 0, \tag{3.3.8}$$

then $d^2\alpha/dt^2 > 0$, and we have a minimum.

Similarly, from the theory of quadratic forms, if the determinant of (3.3.7) is negative and its border-preserving principal first minors (of order 3), such as (3.3.8), are positive, then $d^2\alpha/dt^2 < 0$, and we have a maximum.

We generalize to the sufficient conditions for a function $f(x_1, x_2, \ldots, x_n)$ of n ($\geqslant 3$) variables subject to the constraint $g(x_1, x_2, \ldots, x_n) = 0$ to be a local maximum at the point $(a_1, a_2, \ldots, a_n)$:

$$D = \begin{vmatrix} f_{11}+\lambda g_{11} & \cdots & f_{1n}+\lambda g_{1n} & g_1 \\ \vdots & & & \\ f_{1n}+\lambda g_{1n} & \cdots & f_{nn}+\lambda g_{nn} & g_n \\ g_1 & \cdots & g_n & 0 \end{vmatrix} \begin{cases} > 0 & \text{if } n \text{ is even,} \\ < 0 & \text{if } n \text{ is odd,} \end{cases} \tag{3.3.9}$$

[2]A good, elementary discussion is to be found in Chiang (1974), pp. 385–90.

and the sequence of border-preserving principal minors of the above determinant D alternate in sign, *i.e.*, they are negative if they are of even order and are positive if they are of odd order.

The sufficient conditions for a local minimum are:

$$D = \begin{vmatrix} f_{11}+\lambda g_{11} & \cdots & f_{1n}+\lambda g_{1n} & g_1 \\ \vdots & & & \\ f_{1n}+\lambda g_{1n} & \cdots & f_{nn}+\lambda g_{nn} & g_n \\ g_1 & \cdots & g_n & 0 \end{vmatrix} < 0 \qquad (3.3.10)$$

and its border-preserving principal minors of all orders ($\geqslant 3$) are negative.

Suppose we are interested in the local extremum of a function $f(x_1, x_2, \ldots, x_n)$ of n variables subject to $m = 2$ conditions

$$g^1(x_1, x_2, \ldots, x_n) = 0$$

and

$$g^2(x_1, x_2, \ldots, x_n) = 0.$$

The related determinant is:

$$D = \begin{vmatrix} f_{11}+\lambda_1 g_{11}^1+\lambda_2 g_{11}^2 & \cdots & f_{1n}+\lambda_1 g_{1n}^1+\lambda_2 g_{1n}^2 & g_1^1 & g_1^2 \\ \vdots & & & & \\ f_{1n}+\lambda_1 g_{1n}^1+\lambda_2 g_{1n}^2 & \cdots & f_{nn}+\lambda_1 g_{nn}^1+\lambda_2 g_{nn}^2 & g_n^1 & g_n^2 \\ g_1^1 & \cdots & g_n^1 & 0 & 0 \\ g_1^2 & \cdots & g_n^2 & 0 & 0 \end{vmatrix}. \qquad (3.3.11)$$

A similar determinant of order $m+n$ exists if there are m conditions $g^j(x_1, x_2, \ldots, x_n) = 0, j = 1, 2, \ldots, m$. Let us call the determinant the *restricted Hessian of f*. We quote without proof the sufficient conditions when the function $f(x_1, x_2, \ldots, x_n)$ is subject to m ($<n$) ($n \geqslant 3$) conditions $g^j(x_1, x_2, \ldots, x_n) = 0, j = 1, 2, \ldots, m$, and when the equations given in (3.2.4) are satisfied.

Theorem 3.3.1. *For a local minimum, they are*:

The restricted Hessian and the border-preserving principal minors of all orders $\geqslant 3$ have the sign $(-1)^m$ at the point $(a_1, a_2 \cdots a_n)$.

For a local maximum, they are:

The restricted Hessian has the sign $(-1)^n$ and each border-preserving principal minor of order $k \geqslant 3$ has the sign $(-1)^{k+m}$. ◻

It may be of interest to have a brief and nonrigorous look at some of the reasons underlying the conditions cited above.

We have looked at minors and have restricted ourselves to border-preserving principal minors. The reason is we are eliminating a variable each time. When we delete the third row and the third column from the determinant of (3.3.6), for example, we eliminate the variable u_3 (let $u_3=0$) to determine the sign of the remaining quadratic form $F(u_1,u_2,u_3)$ of (3.3.5). In the case of a minimum, for instance, we want the quadratic form $F(u_1,u_2,u_3)$ to be positive definite everywhere along the curve $p_1u_1+p_2u_2+p_3u_3=0$. This is to be true even when we set $u_3=0$, that is, F is to be positive at $p_1u_1+p_2u_2=u_3=0$, and the condition for this is the second half of (3.3.6).

Next, we specified the smallest border-preserving principal minor to be at least of order 3 when there is only one constraint $g=0$. Consider a border-preserving principal minor of order 2, such as

$$\begin{vmatrix} f_{11}+\lambda g_{11} & g_1 \\ g_1 & 0 \end{vmatrix} = -g_1^2,$$

which is obviously nonpositive. If there are two constraints, that is, if $m=2$, then the smallest meaningful minor is of order 4, since a minor of order 3 such as

$$\begin{vmatrix} f_{11}+\lambda_1 g_{11}^1+\lambda_2 g_{11}^2 & g_1^1 & g_1^2 \\ g_1^1 & 0 & 0 \\ g_1^2 & 0 & 0 \end{vmatrix}$$

automatically $=0$. Similarly, the order of the smallest such minor increases as m increases.

Finally, we observe that if $f(a_1,a_2,\ldots,a_n)$ is a maximum, then $-f(a_1, a_2,\ldots,a_n)$ is a minimum. Thus it is not necessary to discuss the conditions for both the maximum and the minimum. We can deduce the conditions for one from the other. As an illustration, consider the sufficient conditions for a function of $n=7$ variables subject to $m=2$ constraints to attain a local maximum at $(a_1,a_2,\ldots,a_7)$. This is the same as considering the sufficient conditions for $-f(a_1,\ldots,a_7)$ to be a local minimum. According to Theorem 3.3.1, the restricted Hessian of $-f$ and its border-preserving principal minors are positive. Since $g^1(a_1,\ldots,a_7)=0$ and $g^2(a_1,\ldots,a_7)=0$, $g^1=-g^1$ and $g^2=-g^2$ at $(a_1,\ldots,a_n)$, and the restricted Hessian of $-f$ can be written as

$$\begin{vmatrix} -f_{11}-\lambda_1 g_{11}^1-\lambda_2 g_{11}^2 & \cdots & -f_{17}-\lambda_1 g_{17}^1-\lambda_2 g_{17}^2 & -g_1^1 & -g_1^2 \\ \vdots & & & & \\ -f_{17}-\lambda_1 g_{17}^1-\lambda_2 g_{17}^2 & \cdots & -f_{77}-\lambda_1 g_{77}^1-\lambda_2 g_{77}^2 & -g_7^1 & -g_7^2 \\ -g_1^1 & \cdots & -g_7^1 & 0 & 0 \\ -g_1^2 & \cdots & -g_7^2 & 0 & 0 \end{vmatrix}.$$

$$(3.3.12)$$

Table 3.3.1.

	n odd m odd	n odd m even	n even m odd	n even m even
Minimum	$D<0$	$D>0$	$D<0$	$D>0$
Maximum	$D<0$	$D<0$	$D>0$	$D>0$

This determinant is of order 9 and is positive. If we now switch from $-f$ to f, we multiply each row of (3.3.12) by -1 and obtain the restricted Hessian for f. Since we multiply by an odd number of -1's, the sign of this restricted Hessian for a maximum for f becomes negative. The first principal minor of (3.3.12) is of order 8 and is still positive by Theorem 3.3.1. Now the first principal minor of the restricted Hessian of f, being also of order 8, can be obtained from the principal minor of (3.3.12) by multiplying the eight rows by -1. Hence it too is positive in sign.

We may summarize the rules for the signs of the determinants of Theorem 3.3.1 in Table 3.3.1 ($n \geqslant 3$).

3.4 Examples

Example 1. To find the maximum and minimum distances from the origin to the ellipse

$$g(x,y) = 2x^2 + 3xy + 2y^2 - 4 = 0,$$

we determine the extremal values of the function

$$f(x,y) = x^2 + y^2,$$

subject to the condition $g(x,y) = 0$. Using (3.2.3), we have

$$2x + \lambda(4x + 3y) = 0,$$

$$2y + \lambda(3x + 4y) = 0,$$

$$2x^2 + 3xy + 2y^2 - 4 = 0.$$

From the first two equations we get

$$3\lambda(x^2 - y^2) = 0,$$

or $y = \pm x$. Substituting these values in the third equation gives

$$x^2 = \frac{4}{7}, \qquad x = \frac{2}{\sqrt{7}}, \qquad y = \frac{2}{\sqrt{7}}, \quad \text{or} \quad x = \frac{-2}{\sqrt{7}}, \qquad y = \frac{-2}{\sqrt{7}},$$

with $\lambda = -2/7$; and

$$x^2 = 4, \qquad x = 2, \qquad y = -2, \quad \text{or} \quad x = -2, \qquad y = 2, \quad \text{with } \lambda = -2.$$

Obviously, the first value gives a minimum, with $f = x^2 + y^2 = 8/7$, and the second gives a maximum, with $f = 8$.

In most situations, as in this one, the nature of the problem is such that once an extremal value is found, one can determine whether it is a maximum or a minimum by direct substitution.

However, let us try the sufficiency conditions. We have:

$$f_{11}=f_{22}=2, \qquad f_{12}=0,$$
$$g_1=4x+3y,$$
$$g_2=3x+4y,$$
$$g_{11}=g_{22}=4, \qquad g_{12}=3.$$

At $x=2, y=-2$, or $x=-2, y=2$, $\lambda=-2$, the determinant

$$\begin{vmatrix} f_{11}+\lambda g_{11} & f_{12}+\lambda g_{12} \\ f_{12}+\lambda g_{12} & f_{22}+\lambda g_{22} \end{vmatrix} = \begin{vmatrix} -6 & -6 \\ -6 & -6 \end{vmatrix} = 0,$$

so that we cannot use the condition stated for E of (3.3.3).

The determinant of (3.3.9) is

$$\begin{vmatrix} -6 & -6 & 2 \\ -6 & -6 & -2 \\ 2 & -2 & 0 \end{vmatrix} = 96>0 \quad \text{for } x=2, y=-2,$$

$$\begin{vmatrix} -6 & -6 & -2 \\ -6 & -6 & 2 \\ -2 & 2 & 0 \end{vmatrix} = 96>0 \quad \text{for } x=-2, y=2.$$

Therefore the points $(2,-2)$ and $(-2,2)$ give us the maxima.

At $x=2/\sqrt{7}$, $y=2/\sqrt{7}$, or $x=-2/\sqrt{7}$, $y=-2/\sqrt{7}$, $\lambda=-2/7$, we have

$$\begin{vmatrix} \dfrac{6}{7} & -\dfrac{6}{7} & \dfrac{14}{\sqrt{7}} \\ -\dfrac{6}{7} & \dfrac{6}{7} & \dfrac{14}{\sqrt{7}} \\ \dfrac{14}{\sqrt{7}} & \dfrac{14}{\sqrt{7}} & 0 \end{vmatrix} \quad \text{or} \quad \begin{vmatrix} \dfrac{6}{7} & -\dfrac{6}{7} & -\dfrac{14}{\sqrt{7}} \\ -\dfrac{6}{7} & \dfrac{6}{7} & -\dfrac{14}{\sqrt{7}} \\ -\dfrac{14}{\sqrt{7}} & -\dfrac{14}{\sqrt{7}} & 0 \end{vmatrix},$$

both of which are negative. The points $(2/\sqrt{7},2/\sqrt{7})$ and $(-2/\sqrt{7}, -2/\sqrt{7})$ give the minima.

Example 2. In earlier examples we considered the problem of finding the most profitable level of production of an entrepreneur who produces only one good. Now suppose he produces two goods or outputs Q_1 and Q_2 with a single input X. Let q_1, q_2, and x be the respective quantities of Q_1, Q_2, and X, and let

$$x=h(q_1, q_2)$$

be the *production function*, i.e., the cost of production in terms of X is a function h of the quantities of the two outputs.

A problem that is frequently encountered in economics is one in which the entrepreneur wishes to maximize his revenue

$$R = p_1 q_1 + p_2 q_2$$

with a given level of input $x = x_0$. p_1 and p_2 are the unit prices of Q_1 and Q_2 respectively. Each is assumed to be given under purely competitive markets. We further assume that Q_1 and Q_2 are linearly related and that the production of one will necessarily be at the expense of the other.

In our terminology, when p_1 and p_2 are fixed, $R = f(q_1, q_2)$, and we may let the constraint equation be

$$g(q_1, q_2) = x_0 - h(q_1, q_2) = 0.$$

We form the Lagrangian

$$L = p_1 q_1 + p_2 q_2 + \lambda [x_0 - h(q_1, q_2)]$$

and take the first partial derivatives to get

$$L_1 = p_1 - \lambda h_1 = 0,$$
$$L_2 = p_2 - \lambda h_2 = 0,$$
$$L_\lambda = x_0 - h(q_1, q_2) = 0.$$

From these we get the first-order condition

$$\lambda = \frac{p_1}{h_1} = \frac{p_2}{h_2}.$$

The rate at which Q_1 (Q_2) is sacrificed in order to produce more Q_2 (Q_1) with the same input X is called the *rate of product transformation* (RPT):

$$\text{RPT} = -\frac{dq_2}{dq_1}.$$

From the total differential

$$dx = 0 = h_1 dq_1 + h_2 dq_2,$$

we have

$$\text{RPT} = -\frac{dq_2}{dq_1} = \frac{h_1}{h_2} = \frac{p_1}{p_2} = \frac{\dfrac{\partial q_2}{\partial x}}{\dfrac{\partial q_1}{\partial x}},$$

where we use the inverse function rule

$$\frac{\partial q_1}{\partial x} = \frac{1}{h_1} \quad \text{and} \quad \frac{\partial q_2}{\partial x} = \frac{1}{h_2}.$$

Also,

$$\lambda = p_1 \frac{\partial q_1}{\partial x} = p_2 \frac{\partial q_2}{\partial x}.$$

That is to say, the value of the marginal productivity of X in the production of Q_1 or Q_2 is equal to λ.

For a maximum, the determinant of (3.3.9) is positive:

$$\begin{vmatrix} -\lambda h_{11} & -\lambda h_{12} & -h_1 \\ -\lambda h_{12} & -\lambda h_{22} & -h_2 \\ -h_1 & -h_2 & 0 \end{vmatrix} > 0.$$

Expanding,

$$\lambda\left(h_{11}h_2^2 - 2h_{12}h_1 h_2 + h_{22}h_1^2\right) > 0.$$

Since the nature of the problem requires $\lambda > 0$ and $h_i > 0$, $i = 1, 2$, we have

$$h_{11}h_2^2 - 2h_{12}h_1 h_2 + h_{22}h_1^2 > 0.$$

From the above, the total derivative of RPT is its rate of change and is

$$-\frac{d^2 q_2}{dq_1^2} = \frac{1}{h_2^3}\left(h_{11}h_2^2 - 2h_{12}h_1 h_2 + h_{22}h_1^2\right).$$

Therefore, the sufficiency condition requires that the product transformation curve have an increasing RPT.

Example 3. As a dual problem to the one given above, suppose the entrepreneur wishes to minimize the amount of X needed to obtain a specified revenue R. That is, he wishes to minimize

$$x = h(q_1, q_2)$$

subject to

$$R = r_0 = p_1 q_1 + p_2 q_2.$$

Here we form the Lagrangian

$$G = h(q_1, q_2) + \mu(r_0 - p_1 q_1 - p_2 q_2),$$

take the first partial derivatives to get, again,

$$\mu = \frac{h_1}{p_1} = \frac{h_2}{p_2}$$

and

$$\frac{h_1}{h_2} = \frac{p_1}{p_2}.$$

The sufficiency condition requires that

$$\begin{vmatrix} h_{11} & h_{12} & -p_1 \\ h_{12} & h_{22} & -p_2 \\ -p_1 & -p_2 & 0 \end{vmatrix} < 0.$$

If we multiply each row by $-\mu$, since $\mu > 0$, we get

$$\begin{vmatrix} -\mu h_{11} & -\mu h_{12} & \mu p_1 \\ -\mu h_{12} & -\mu h_{22} & \mu p_2 \\ \mu p_1 & \mu p_2 & 0 \end{vmatrix} = \begin{vmatrix} -\mu h_{11} & -\mu h_{12} & h_1 \\ -\mu h_{12} & -\mu h_{22} & h_2 \\ h_1 & h_2 & 0 \end{vmatrix} > 0.$$

This determinant, of course, is the same as

$$\begin{vmatrix} -\mu h_{11} & -\mu h_{12} & -h_1 \\ -\mu h_{12} & -\mu h_{22} & -h_2 \\ -h_1 & -h_2 & 0 \end{vmatrix}$$

after multiplying the third row and third column by -1. Comparing this with the determinant in the previous example, we see that the two problems are essentially the same, or one is the dual of the other. We notice that μ is the reciprocal of λ of the previous example. In fact, the total differential of

$$R = p_1 q_1 + p_2 q_2$$

is

$$dR = p_1 dq_1 + p_2 dq_2 = \lambda(h_1 dq_1 + h_2 dq_2)$$

by substituting $p_1 = \lambda h_1$ and $p_2 = \lambda h_2$ from the last example. Since $x = h(q_1, q_2)$ and

$$dx = h_1 dq_1 + h_2 dq_2,$$

we get

$$\frac{dR}{dx} = \frac{\lambda(h_1 dq_1 + h_2 dq_2)}{h_1 dq_1 + h_2 dq_2} = \lambda.$$

That is to say, λ is the derivative of the revenue R with respect to x when prices are held constant. The multiplier μ of this example, being the reciprocal of λ, is the derivative of x with respect to R. When the sufficiency condition is satisfied, if $R = r_0$ is the maximum revenue the entrepreneur can achieve with a given level of input x_0, x_0 is the minimum amount of input needed to obtain a specified revenue r_0.

Example 4. A similar dual problem exists when the entrepreneur wishes to maximize his output Q subject to a cost constraint $C = c_0$, or to minimize the cost C of producing a prescribed level of output $Q = q_0$. Let

$$q = f(x_1, x_2),$$

where the lower case letters again represent the values of Q, X_1, and X_2 respectively, and where X_1 and X_2 are two variable inputs. Suppose, further, that the cost of production is

$$C = r_1 x_1 + r_2 x_2 + a,$$

where r_1 and r_2 are the prices of X_1 and X_2 and a is the price of any fixed inputs also needed for the production of Q. For a given

$$C = c_0 = r_1 x_1 + r_2 x_2 + a,$$

the entrepreneur wishes to maximize q. We form the Lagrangian

$$M = f(x_1, x_2) + \lambda(c_0 - r_1 x_1 - r_2 x_2 - a),$$

equate the first partial derivatives to zero, and solve to obtain:

$$\frac{f_1}{f_2} = \frac{r_1}{r_2},$$

which means that the ratio of the marginal productivities of X_1 and X_2 must be equal to the ratio of their prices. Also,

$$\lambda = \frac{f_1}{r_1} = \frac{f_2}{r_2},$$

which states that λ is equal to the contribution of the money spent on each unit of input to output. Finally,

$$dC = r_1 dx_1 + r_2 dx_2$$

$$= \frac{1}{\lambda}(f_1 dx_1 + f_2 dx_2).$$

But $dq = f_1 dx_1 + f_2 dx_2$. So

$$\frac{dq}{dC} = \lambda,$$

that is, λ is the derivative of output with respect to cost with fixed prices and variable inputs.

The sufficiency condition requires that

$$\begin{vmatrix} f_{11} & f_{12} & -r_1 \\ f_{12} & f_{22} & -r_2 \\ -r_1 & -r_2 & 0 \end{vmatrix} > 0.$$

Now consider the dual situation when the entrepreneur wishes to minimize the cost of producing

$$q = q_0 = f(x_1, x_2).$$

Form the Lagrangian

$$N = r_1 x_1 + r_2 x_2 + a + \mu[q_0 - f(x_1, x_2)];$$

proceeding as before, we get

$$\frac{1}{\mu} = \frac{f_1}{r_1} = \frac{f_2}{r_2}$$

and

$$\frac{f_1}{f_2} = \frac{r_1}{r_2}.$$

Thus, μ is the derivative of cost with respect to output level.

Finally, the sufficiency condition for this minimum problem is

$$\begin{vmatrix} -\mu f_{11} & -\mu f_{12} & -f_1 \\ -\mu f_{12} & -\mu f_{22} & -f_2 \\ -f_1 & -f_2 & 0 \end{vmatrix} < 0.$$

But this is equivalent to

$$\begin{vmatrix} f_{11} & f_{12} & -r_1 \\ f_{12} & f_{22} & -r_2 \\ -r_1 & -r_2 & 0 \end{vmatrix} > 0,$$

as noted in the previous revenue-input example.

If q_0 is the maximum output which can be produced with a given outlay of c_0 dollars, then c_0 is the minimum cost for which the output q_0 may be produced.

Example 5. Suppose that the output q of a firm is related to the inputs of labor l and capital k by the function $q = 120 l^{1/2} k^{1/3}$. We wish to find the cost-minimizing input levels for a given output level $\bar{q}$ if the price (rental) of capital is r and the wage rate is w. The total cost C is $C = wl + rk$, which is to be minimized, subject to $g(l,k) = 120 l^{1/2} k^{1/3} - \bar{q} = 0$. The Lagrangian is

$$L = rk + wl + \mu(\bar{q} - 120 l^{1/2} k^{1/3}).$$

At minimal cost, we must have

$$\frac{\partial L}{\partial k} = r - 40 \mu l^{1/2} k^{-2/3} = 0, \qquad \frac{\partial L}{\partial l} = w - 60 \mu l^{-1/2} k^{1/3} = 0,$$

$$\frac{\partial L}{\partial \mu} = \bar{q} - 120 l^{1/2} k^{1/3} = 0.$$

Since r, w, l, and k are all assumed to be positive by the nature of the problem, μ must be positive. Also,

$$\frac{w}{r} = \frac{3}{2} \frac{k}{l}, \quad i.e., \quad k = \frac{2}{3} \frac{wl}{r}.$$

From

$$\bar{q} = 120 l^{1/2} \left(\frac{2}{3} \frac{wl}{r} \right)^{1/3} = 120 \left(\frac{3}{2} \frac{kr}{w} \right)^{1/2} k^{1/3},$$

we obtain the optimal input levels:

$$l = \left(\frac{\bar{q}}{120} \right)^{6/5} \left(\frac{3r}{2w} \right)^{2/5}, \qquad k = \left(\frac{\bar{q}}{120} \right)^{6/5} \left(\frac{2w}{3r} \right)^{3/5}.$$

To ascertain that we have found a minimum, we check the second order conditions:

$$\frac{\partial^2 L}{\partial k^2} = \frac{80}{3}\,\mu l^{1/2}k^{-5/3} > 0, \qquad \frac{\partial^2 L}{\partial k \partial l} = -20\mu l^{-1/2}k^{-2/3} < 0,$$

$$\frac{\partial^2 L}{\partial l^2} = 30\mu k^{1/3}l^{-3/2} > 0, \qquad \frac{\partial^2 L}{\partial k^2}\cdot\frac{\partial^2 L}{\partial l^2} - \left(\frac{\partial^2 L}{\partial k \partial l}\right)^2 = 400\mu^2 l^{-1}k^{-4/3} > 0,$$

so we have a minimum. The same problem may be phrased as one maximizing the profit π, where, for prescribed price p and output q,

$$\pi = pq - wl - rk,$$

subject to

$$q = 120l^{1/2}k^{1/3} = \bar{q} = \text{constant}.$$

The Lagrangian now is

$$M = pq - wl - rk + \lambda(\bar{q} - 120l^{1/2}k^{1/3}).$$

The first- and second-order derivatives of M with respect to k and l and λ will be almost identical to those obtained above, with μ replaced by $-\lambda$. We notice that λ is negative, and that

$$\frac{\partial^2 M}{\partial l^2} = 30\lambda l^{-3/2}k^{1/3} < 0, \qquad \frac{\partial^2 M}{\partial l \partial k} = 20\lambda l^{1/2}k^{-2/3},$$

$$\frac{\partial^2 M}{\partial k^2} = \frac{80}{3}\lambda l^{1/2}k^{-5/3} < 0,$$

$$\frac{\partial^2 M}{\partial l^2}\cdot\frac{\partial^2 M}{\partial k^2} - \left(\frac{\partial^2 M}{\partial l \partial k}\right)^2 = 400\lambda^2 l^{-1}k^{-1/3} > 0,$$

so that the values solved for l and k will give us maximum profit.[3]

Example 6. Another application is to be found in the maximization of the utility function of a consumer. Suppose that there are three commodities Q_1, Q_2, and Q_3 with given prices p_1, p_2, and p_3 respectively, and that the consumer has a certain given amount of money m (called income for convenience) to spend on the purchase of these three goods. The consumer wishes to maximize the function

$$U = u(q_1, q_2, q_3),$$

which indicates his satisfaction when he buys q_i units of the goods Q_i, $i = 1, 2, 3$, subject to the constraint

$$m = p_1 q_1 + p_2 q_2 + p_3 q_3.$$

---

[3]Question for economists: If the production function used should be of constant returns to scale Cobb–Douglas type, would the second-order condition still be satisfied under profit maximization when the input and output prices are given?

We form the Lagrangian

$$u(q_1, q_2, q_3) + \lambda(m - p_1 q_1 - p_2 q_2 - p_3 q_3),$$

and equate the first partial derivatives to zero as we have done. One result
is

$$\frac{\partial u}{\partial q_i} = \lambda p_i, \qquad i = 1, 2, 3.$$

That is to say, the marginal utility of the ith commodity is proportional to
the ith price at the optimal point, and λ is the constant of proportionality.
Another way to write λ is, since p_i is a constant,

$$\lambda = \frac{\partial u}{\partial (p_i q_i)}, \qquad i = 1, 2, 3.$$

At the optimal point, the utility of the consumer is increased by the same λ
units when his money income m is increased for him to spend on any of
the three goods. λ is often called the marginal utility of income.

As a specific example, suppose

$$u(q_1, q_2, q_3) = -q_1^2 - q_2^2 - 2q_3^2 + 10q_1 + 20q_2 + 40q_3,$$

subject to

$$g(q_1, q_2, q_3) = m - p_1 q_1 - p_2 q_2 - p_3 q_3 = 0.$$

We take the partial derivatives of

$$-q_1^2 - q_2^2 - 2q_3^2 + 10q_1 + 20q_2 + 40q_3 + \lambda(m - p_1 q_1 - p_2 q_2 - p_3 q_3)$$

with respect to the q_i and to λ and get

$$2q_1 + \lambda p_1 = 10$$

$$2q_2 + \lambda p_2 = 20$$

$$4q_3 + \lambda p_3 = 40$$

$$p_1 q_1 + p_2 q_2 + p_3 q_3 = m.$$

We solve for the q_i and λ. For example, we find

$$q_1 = \frac{2p_1 m + 10p_2^2 + 5p_3^2 - 20p_1 p_2 - 20p_1 p_3}{2p_1^2 + 2p_2^2 + p_3^2}$$

and

$$\lambda = \frac{4(5p_1 + 10p_2 + 10p_3 - m)}{2p_1^2 + 2p_2^2 + p_3^2}.$$

The second partial derivatives are

$$u_{11} = -2, \qquad u_{22} = -2, \qquad u_{33} = -4, \qquad u_{ij} = 0 \quad \text{for } i \neq j,$$

$$g_{ii} = g_{ij} = 0 \quad \text{for all } i, j = 1, 2, 3.$$

The determinant

$$\begin{vmatrix} -2 & 0 & 0 & -p_1 \\ 0 & -2 & 0 & -p_2 \\ 0 & 0 & -4 & -p_3 \\ -p_1 & -p_2 & -p_3 & 0 \end{vmatrix} = -4\left(2p_1^2 + 2p_2^2 + p_3^2\right) < 0$$

and its principal minor

$$\begin{vmatrix} -2 & 0 & -p_2 \\ 0 & -4 & -p_3 \\ -p_2 & -p_3 & 0 \end{vmatrix} = 4p_2^2 + 2p_3^2 > 0,$$

as stated in the condition given by (3.3.9).

q_1, as a function of p_1, p_2, p_3, and m as derived above, is known as the *ordinary demand function*. Another demand function, known as the *compensated demand function*, can be derived on the assumption that the government taxes or subsidizes a consumer so as to leave his utility level unchanged after a price change. The compensated demand functions are obtained by *minimizing* the consumer's expenditures subject to a fixed level of *utility* u_0. Using again the utility function

$$u = -q_1^2 - q_2^2 - 2q_3^2 + 10q_1 + 20q_2 + 40q_3 = u_0 = \text{constant},$$

the Lagrangian is

$$L = p_1 q_1 + p_2 q_2 + p_3 q_3 + \mu\left[u_0 - \left(-q_1^2 - q_2^2 - 2q_3^2 + 10q_1 + 20q_2 + 40q_2\right)\right].$$

Take the partial derivatives with respect to the q_i and to μ and get

$$p_1 + \mu(2q_1 - 10) = 0,$$
$$p_2 + \mu(4q_2 - 20) = 0,$$
$$p_3 + \mu(4q_3 - 40) = 0,$$
$$u_0 = -q_1^2 - q_2^2 - 2q_3^2 + 10q_1 + 20q_2 + 40q_3,$$

from which we may solve for the q_i and μ.

Example 7. Suppose a firm is investing in n distinct facilities for manufacturing a homogeneous product. These facilities may be plants, shifts, or units of equipment. Let $i = 1, 2, \ldots, n$, q_i be the output of the ith facility and let $c_i = c_i(q_i)$ be the current cost of the ith facility. The total cost is

$$c = \sum_{i=1}^{n} c_i(q_i).$$

The firm wishes to minimize its total cost, subject to the requirement that the total output q must be

$$q = \sum_{i=1}^{n} q_i.$$

Form the Lagrangian

$$L = \sum_{i=1}^{n} c_i(q_i) + \lambda\left(q - \sum_{i=1}^{n} q_i\right)$$

and take partial derivatives as usual. The first partial derivatives yield the result that

$$\lambda = \frac{dc_1}{dq_1} = \frac{dc_2}{dq_2} = \cdots = \frac{dc_n}{dq_n}.$$

These conditions $(dc_i/dq_i) - \lambda = 0$, together with the constraint

$$c = \sum_{i=1}^{n} c_i(q_i),$$

determine each optimal $q_i = q_i^o$ as a function of q, that is,

$$q_i^o = h_i(q), \qquad i = 1, 2, \ldots, n.$$

Therefore,

$$c = \Sigma c_i\left[h_i(q)\right]$$

under optimal allocation of costs, and λ is the firm's marginal current total cost,

$$\lambda = \frac{dc}{dq}.$$

The sufficient conditions require that the determinant

$$\begin{vmatrix} c_1'' & 0 & 0 & \cdots & 1 \\ 0 & c_2'' & 0 & \cdots & 1 \\ 0 & 0 & c_3'' & \cdots & 1 \\ \vdots & & & c_n'' & \\ 1 & 1 & 1 & \cdots & 0 \end{vmatrix},$$

where

$$c_i'' = \frac{d^2 c_i}{dq_i^2}, \qquad i = 1, \ldots, n,$$

and its border-preserving principal minors be negative. These determinants, when expanded, yield

$$\left(\frac{1}{c_1''} + \frac{1}{c_2''} + \ldots + \frac{1}{c_n''}\right) c_1'' c_2'' \ldots c_n'' > 0,$$

$$\vdots$$

$$\left(\frac{1}{c_1''} + \frac{1}{c_2''} + \frac{1}{c_3''}\right) c_1'' c_2'' c_3'' > 0,$$

$$c_1'' + c_2'' > 0.$$

Since we may take any border-preserving principal minor, these conditions apply to any two, three, or more of the n facilities, not necessarily restricted to the first or second or third.

A number of interesting observations may be made as a result of these conditions. For example, from

$$c_i'' + c_j'' > 0, \qquad i \neq j,$$

we see that under optimal allocation at most one facility may be operated at declining marginal cost. If, for example, the first facility is operated at declining marginal cost, then $-c_1'' < c_j''$ or $|c_1''| < c_j''$ for all $j \neq 1$. The rate of decline of the marginal cost in the first facility must be smaller in absolute value than the rate of increase of the marginal cost of every other facility.

Moreover, if $c_1'' < 0$, and we know from above that $c_j'' > 0$ for all $j \neq 1$, then $c_1'' c_2'', \ldots, c_n'' < 0$. In order to have

$$\left(\frac{1}{c_1''} + \frac{1}{c_2''} + \ldots + \frac{1}{c_n''} \right) c_1'' c_2'' \ldots c_n'' > 0,$$

we must have

$$\frac{1}{c_1''} + \sum_{j=2}^{n} \frac{1}{c_j''} < 0,$$

that is,

$$|c_1''| < \frac{1}{\displaystyle\sum_{j=2}^{n} \frac{1}{c_j''}}.$$

Example 8. Imagine an economy in which there is a point-rationing system under which each consumer is given a certain number of points. For each good, there is a money price as well as point price, so that the consumer has to pay both money and points to purchase a commodity. It is illegal to transfer points from one person to another.

Suppose there are three goods Q_1, Q_2, and Q_3, with their respective money and point prices m_i and p_i, and that q_i are the quantities of the goods purchased, $i = 1, 2, 3$. The consumer wishes to maximize his utility function

$$u = u(q_1, q_2, q_3)$$

subject to the money and point budget constraints:

$$m_1 q_1 + m_2 q_2 + m_3 q_3 = \bar{m} = \text{constant};$$

$$p_1 q_1 + p_2 q_2 + p_3 q_3 = \bar{p} = \text{constant}.$$

The Lagrangian is

$$L = u(q_1, q_2, q_3) + \lambda_1 \left(\bar{m} - \sum_{i=1}^{3} m_i q_i \right) + \lambda_2 \left(\bar{p} - \sum_{i=1}^{3} p_i q_i \right).$$

The partial derivatives, equated to 0, are

$$\frac{\partial L}{\partial q_i} = \frac{\partial u}{\partial q_i} - \lambda_1 m_i - \lambda_2 p_i = 0, \qquad i = 1, 2, 3,$$

and the two constraint equations. These first-order conditions show that

$$\frac{\dfrac{\partial u}{\partial q_i}}{\dfrac{\partial u}{\partial q_j}} = \frac{\lambda_1 m_i + \lambda_2 p_i}{\lambda_1 m_j + \lambda_2 p_j}, \qquad i = 1, 2, 3, \ \ j = 1, 2, 3.$$

The marginal rate of substitution between goods Q_i and Q_j is equal to the ratios of the weighted averages of the money and point prices.

Exercises

3.1 Let q_1 and q_2 be the quantities of two goods Q_1 and Q_2 selling at unit prices \$4 and \$10 respectively. If a consumer has \$200 and his utility function is $u = q_1 q_2$, how many units of each good should he buy to maximize his utility? Is the second-order condition satisfied?

3.2 Suppose the volume of sales S of a product is a function of the number of newspaper ads x and number of minutes of TV time y: $S = 12xy - x^2 - 3y^2$. Each newspaper ad or each minute on TV costs \$100. How should the firm allocate \$1,600 between these two advertising media in order to achieve maximum sales?

3.3 100 persons consume two goods. Each person has a fixed income of \$1,000 and the same utility function

$$u = 15q_1 + 10q_2 + q_1 q_2.$$

Let the unit price of good Q_2 be \$5. Express the aggregate demand for good Q_1 as a function of p_1, the unit price of Q_1. What is the slope of this aggregate demand function?

3.4 Referring to Example 8 in Section 3.3, assume that the consumer may use either money or points to buy the goods. (a) Derive the first- and second-order conditions for a maximum. (b) Show that λ_1 is the marginal utility of income with m_i, p_i and $\bar{p}$ constant, and q_i variable, while λ_2 is the marginal utility of point allotment with m_i, p_i, and $\bar{m}$ constant. (c) Find a sufficient condition which will allow the consumer's purchases to be unaffected by the imposition of points.

3.5 A consumer produces certain goods and services Y at home, part of which he consumes. Let $y =$ units of Y which he consumes, $T =$ hours of work (or leisure), $t =$ hours spent in producing Y, $M =$ money income, $w =$ wage rate he earns outside of home, $p =$ unit price of Y. $Y = F(t)$ is the production function. His income is $M = p[F(t) - y] + w(T - t)$.

If his utility depends on T, M and y, that is, $u = u(T, M, y)$, derive the conditions that u may be maximized. Show that at optimal t,

$$w = p \frac{dF}{dt} = -\frac{\dfrac{\partial u}{\partial T}}{\dfrac{\partial u}{\partial M}} \quad \text{and} \quad p = \frac{\dfrac{\partial u}{\partial y}}{\dfrac{\partial u}{\partial M}}.$$

3.6 An economy possesses fixed amounts $\bar{K}$ and $\bar{L}$ of two inputs K and L which go into the production of two outputs X and Y. Let K_x and L_x, and K_y and L_y, be the amounts of K and L that go into the productions of X and Y respectively. The production functions for X and Y are

$$X = f(K_x, L_x) \quad \text{and} \quad Y = g(K_y, L_y).$$

The transformation or production possibility curve of the economy gives the maximum output of Y for each given output $\bar{X}$ of X, subject to resource constraints $K_x + K_y = \bar{K}$ and $L_x + L_y = \bar{L}$. Find the conditions for obtaining the transformation curve $Y = Y(X)$ and show that the slope of that curve is the Lagrange multiplier λ. *Hint*: Write $\bar{X}$ as $\bar{X} = f(\bar{K} - K_y, \bar{L} - L_y)$.

3.7 n and k amounts, respectively, of two factors N and K go into the production of a good X whose production function is

$$x = (an^{-\beta} + bk^{-\beta})^{-1/\beta},$$

where a, b, and β are positive constants. The prices of N and K are assumed given as p_n and p_k. Find the conditions which will determine the amounts of N and K to be used to produce $\bar{x}$ units of X at minimum cost.

4

Simultaneous Optima of Several Functions

4.1 Statement of the problem

Up to now we have looked at the extrema of one function $f(x_1, x_2, \ldots, x_n)$ of n variables. In many applied problems, one may be confronted with several functions, say $f_1(x_1, \ldots, x_n), f_2(x_1, \ldots, x_n), \ldots, f_m(x_1, \ldots, x_n)$, with $m \leqslant n$, and wish to maximize them simultaneously. Unless there happens to be a particularly fortunate combination of circumstances, it is usually unlikely that we can find a point $(x_1^o, \ldots, x_n^o)$ such that all, or more than one, of the m functions attain maxima at that point, when each function is taken individually with no regard for the other $m-1$ functions. Nevertheless, it cannot be denied that often in economics, and in other disciplines, situations arise in which we wish to "do as well as we can" for each of several functions. Just what one means by "as well as we can" will depend on the situation.

One approach to the problem, which amounts to converting the situation to one involving one function, is to form a composite function $F = F(f_1, f_2, \ldots, f_m)$, which is a function of the m functions f_i, $i = 1, \ldots, m$, given in some explicit form, and try to maximize F. As an illustration, suppose that there are two sellers of a product which is produced at constant marginal cost and sold at price p, with

$$p = a - b(q_A + q_B),$$

where q_A and q_B are the quantities of the product sold by sellers A and B, and a and b are positive constants. If the cost to each seller is

$$C_i = cq_i + d, \qquad i = A, B,$$

where c and d are positive constants, then the profit for each firm is

$$\pi_i = [a - b(q_A + q_B)]q_i - cq_i - d, \qquad i = A, B.$$

Suppose that these two duopolists are willing to pool their profits together and maximize their joint profit

$$\pi = \pi_A + \pi_B = [a - b(q_A + q_B)](q_A + q_B) - c(q_A + q_B) - 2d.$$

We then must have

$$\frac{\partial \pi}{\partial q_A} = \frac{\partial \pi}{\partial q_B} = \left[a - b(q_A + q_B)\right] - b(q_A + q_B) - c = 0,$$

which yields the set of optimal solutions, with

$$q_A + q_B = \frac{a - c}{2b}.$$

The midpoint of this line segment at

$$q_A = q_B = \frac{a - c}{4b}$$

is the symmetric joint maximum point. It is not possible for either seller to increase his profit by moving to any other point on this line without reducing the profit of his competitor.[1]

In some economic problems, it may be possible to form a composite function F in which

$$F(f_1, \ldots, f_m) = w_1 f_1 + w_2 f_2 + \ldots + w_m f_m,$$

where the w's stand for weights. For instance, certain specified prices can be taken for weights. In general, however, it will not be a simple matter to arrive at a composite function F, let alone to find its maximum.

The more commonly adopted approach involves the notion of *Pareto optimality* and is useful in the studies of welfare economics and of the theory of production, among others. Suppose that, given the values of m functions at a fixed feasible point $\bar{x} = (\bar{x}_1, \bar{x}_2, \ldots, \bar{x}_n)^T$, that is, given the values of $f_1(\bar{x}_1, \ldots, \bar{x}_n), \ldots, f_m(\bar{x}_1, \ldots, \bar{x}_n)$, we wish to know whether we can

---

[1]There is a well-known Cournot solution to this problem. Suppose that each seller believes that any change in the quantity he sells will not affect the quantity his competitor sells, that is, $\partial q_i / \partial q_j = 0$, for $i \neq j$. Under this assumption, seller A should choose his q_A to be

$$q_A = \frac{a - c - bq_B}{2b}$$

by solving the equation (from the necessary condition of Chapter 3)

$$\frac{\partial \pi_A}{\partial q_A} = a - b(q_A + q_B) - bq_A - bq_B \frac{\partial q_B}{\partial q_A} - c = 0.$$

Similarly, for seller B, under the same assumption,

$$q_B = \frac{a - c - bq_A}{2b}.$$

When these two equations are solved together, we arrive at the solution

$$q_A^o = q_B^o = \frac{a - c}{3b}$$

and this pair of quantities (q_A^o, q_B^o) constitutes what is called the *Cournot equilibrium* or *simultaneous maxima for both π_A and π_B.*

move to a different feasible point, say $\hat{x}=(\hat{x}_1,\ldots,\hat{x}_n)^T$ and increase the value of at least one of the functions, while none of the others decreases; that is,

$$f_k(\hat{x}_1,\ldots,\hat{x}_n)>f_k(\bar{x}_1,\ldots,\bar{x}_n) \quad \text{for some } k$$

and

$$f_i(\hat{x}_1,\ldots,\hat{x}_n)<f_i(\bar{x}_1,\ldots,\bar{x}_n) \quad \text{for no } i, i=1,\ldots,m.$$

If such a point $\hat{x}$ exists, then clearly $\hat{x}$ is more desirable than $\bar{x}$ (as long as larger values of f correspond to improvement).

A point $x=(x_1,\ldots,x_n)^T$ is said to be a *Pareto optimal point for the m functions* $f_1(x_1,\ldots,x_n),\ldots,f_m(x_1,\ldots,x_n)$, *subject to constraint*(s) C if (i) x satisfies the constraint(s) C and (ii) there does not exist any point $\hat{x}=(\hat{x}_1,\ldots,\hat{x}_n)^T$ which also satisfies the constraints such that in moving from x to $\hat{x}$ no function decreases while at least one function actually increases. That is, there does not exist $\hat{x}$ satisfying the constraint(s) C such that

$$f_i(\hat{x}_1,\ldots,\hat{x}_n) \geqslant f_i(x_1,\ldots,x_n) \quad \text{for all } i=1,\ldots,m$$

and (4.11)

$$f_k(\hat{x}_1,\ldots,\hat{x}_n)>f_k(x_1,\ldots,x_n) \quad \text{for some } k.$$

There is, in general, an infinite number of Pareto optimal points for a given set of functions. For instance, suppose we have two functions of a single real variable x, say

$$f_1(x)=x \quad \text{and} \quad f_2(x)=-x.$$

Then any point $x=a$ is a Pareto optimal point. If we move to any point which increases f_1, certainly f_2 will decrease.

A point x satisfying the constraint(s) is a *local Pareto optimal point* if there is a neighborhood N of x so that there is no $\hat{x} \in N$ satisfying the constraint(s) such that (4.1.1) holds.

4.2 Graphic Interpretation

We may have a better understanding of the notion of a Pareto optimal point by looking at the so-called Edgeworth–Bowley box diagram.

Suppose there are two inelastically supplied resources, r_1 and r_2, used to produce two goods x and y. The quantities of the two goods produced are functions of the resources:

$$q_x=f_1(r_{x1},r_{x2}),$$
$$q_y=f_2(r_{y1},r_{y2})$$

(4.2.1)

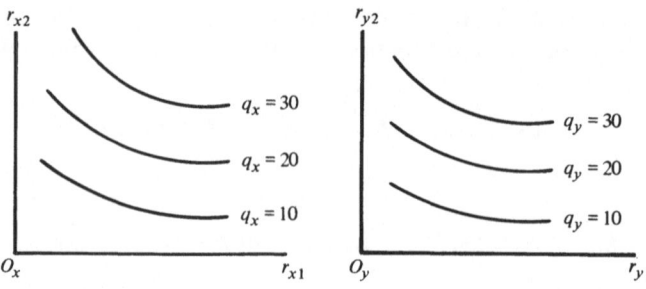

Figure 4.2.1.   **Figure 4.2.2.**

where r_{xi} is the input of r_i for the production of good x, $i=1,2$, and similarly for r_{yi}. Since the resources are inelastically supplied,

$$r_{x1}+r_{y1}=\bar{r}_1 ,$$

$$r_{x2}+r_{y2}=\bar{r}_2 ,$$

(4.2.2)

where $\bar{r}_i$ are fixed available amounts of resource i, $i=1,2$.

These resource and quantity levels may be represented in the following figures in which we have drawn a few sample values for q_x and q_y. The curves are called *isoquants*, that is, on each curve the value of q_x (or q_y) is the same for the combinations of r_{x1} and r_{x2} (or r_{y1} and r_{y2}).

We combine Figures 4.2.1 and 4.2.2 into the Edgeworth–Bowley box in the diagram in Figure 4.2.3, where the dimensions of the box are the given amounts of $\bar{r}_1$ (horizontal) and $\bar{r}_2$ (vertical). The lower left-hand corner is the origin of the axes for product x in Figure 4.2.1, and r_{x1} and r_{x2} are measured from this corner. The upper right-hand corner is the origin of the axes for product y of Figure 4.2.2, turned around diagonally.

Every point in the box, such as point A, represents a set of given values for r_{x1}, r_{x2}, r_{y1}, r_{y2}, q_x and q_y satisfying (4.2.1) and (4.2.2). Let us compare the points A and B. At B, $q_x=10$ and $q_y=20$. It is clearly not as desirable

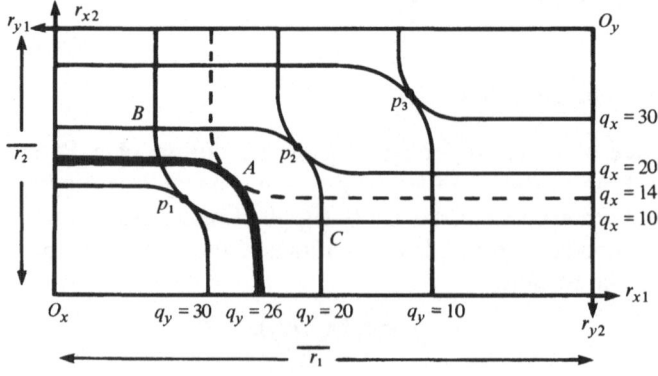

Figure 4.2.3.

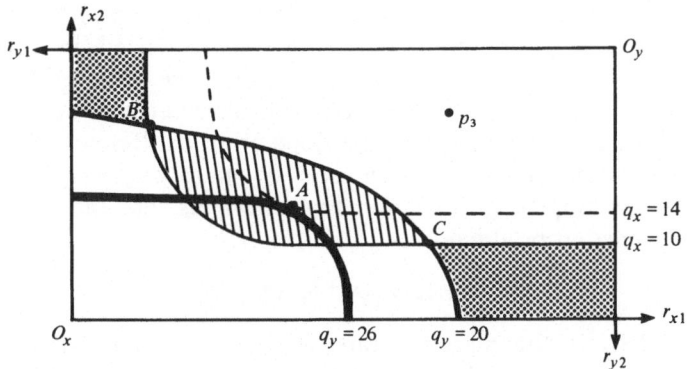

Figure 4.2.4.

as A, where $q_x = 14$ and $q_y = 26$. In Figure 4.2.4, we reproduce that portion
of Figure 4.2.3 which shows the isoquants (the collection of points in which
the quantities are the same) $q_x = 10$ and $q_y = 20$. The point A has already
been shown to be superior to either B or C (where $q_x = 10$ and $q_y = 20$). In
fact, every point in the shaded region is more desirable than B or C, so any
point in that region would represent an increase of both q_x and q_y. That is
to say, the shaded region is the locus of points where both functions f_1 and
f_2 of (4.2.1) may be improved upon if we started with the given curves
$q_x = 10$ and $q_y = 20$. In the two dotted areas outside, both functions will
diminish in value. In the rest of the box, one function will increase at the
expense of the other. For example, at p_3, q_x will be increased to 30, but q_y
will decrease to 10. Thus, starting at points B or C, we may move inward
to improve our positions. Whether we wish to stop at A or at another
point, say p_3, will depend on what relative values of q_x and q_y we desire.
However, if we start at a point which is the point of tangency of two
isoquants, then it is not possible to increase one function without sacrific-
ing the other. Such a point is called a *Pareto efficient point*.[2] For each given
isoquant q_x there is an isoquant q_y which is tangent to it, and *vice versa*,
and the point of tangency is an efficient point. The curve (not drawn)
which includes the locus of all points of tangency is called the *production
curve*.

If, instead of the production of two goods, we are interested in, say, the
utilities of two consumers in the purchase or consumption of two goods, a
similar Edgeworth–Bowley box diagram may be drawn showing the *indifference curves* of the two consumers. Again, the points of tangency of the
two sets of indifference curves are the Pareto optimal points, and the locus
of all these points is called the *contract curve*.

---

[2]The usual terminology in economics uses "Pareto efficient point" in production and reserves
"Pareto optimal point" in a more general context dealing with utility. Either point satisfies
our definition of Pareto optimality.

It will be noticed that these Pareto efficient (or optimal) points on the production curve (or contract curve) are characterized in general by equality of the slopes of the isoquants (or indifference curves). The slope of an isoquant is the *marginal rate of technical substitution between inputs* or *the ratio of marginal products*, and the slope of an indifference curve is the *marginal rate of substitution between goods* or *the ratio of marginal utilities*. Therefore, our graphic presentation has shown us that the condition of Pareto efficiency in production is

$$\text{MRTS}^x_{12} = \text{MRTS}^y_{12}, \tag{4.2.3}$$

where MRTS^x_{12} is the marginal rate of technical substitution between inputs of resources 1 and 2 for product x, and similarly for MRTS^y_{12}.

And the condition of Pareto optimality in the distribution of goods is

$$\text{MRCS}^A_{xy} = \text{MRCS}^B_{xy}, \tag{4.2.4}$$

where MRCS^A_{xy} is the marginal rate of substitution between commodities x and y for consumer A, and similarly for MRCS^B_{xy} for consumer B.

4.3 Necessary conditions for Pareto optimality in production, consumption, and the economy

Let us consider analytically the results which we obtained graphically in the last section. Suppose there are $m \geqslant 2$ producers or firms, $c \geqslant 2$ consumers, n goods or commodities Q_k, $k = 1, 2, \ldots, n$, and r primary factors or resources X_j, $j = 1, 2, \ldots, r$. The production functions are

$$F_h = (q_{h1}, q_{h2}, \ldots, q_{hn}, x_{h1}, \ldots, x_{hr}) = 0, \qquad h = 1, 2, \ldots, m, \tag{4.3.1}$$

and the consumer's utility functions are

$$u_i = u_i(q^*_{i1}, \ldots, q^*_{in}, \bar{x}_{i1} - x^*_{i1}, \ldots, \bar{x}_{ir} - x^*_{ir}), \qquad i = 1, 2, \ldots, c, \tag{4.3.2}$$

where $q_{hk} = $ output of Q_k by the hth producer, $x_{hj} = $ amount of jth resource used by the hth producer, $q^*_{ik} = $ amount of Q_k consumed by the ith consumer, $\bar{x}_{ij} = i$th consumer's endowment of resource X_j, and $x^*_{ij} = $ amount of X_j that consumer i supplies to producers.

We consider a closed economy with no increasing returns to scale. The resources are given fixed. Each consumer accepts the prices p_k of Q_k as given and each producer takes the prices π_j of X_j. We assume:

(i)

$$\sum_{i=1}^{c} x^*_{ij} = \sum_{h=1}^{m} x_{hj} = a_j > 0 \quad \text{for } j = 1, \ldots, r \tag{4.3.3}$$

and

$$\sum_{i=1}^{c} q^*_{ik} = \sum_{h=1}^{m} q_{hk} = b_k > 0 \quad \text{for } k = 1, \ldots, n; \tag{4.3.4}$$

(ii) The utility and production functions are continuously twice differentiable.[3] The first derivatives are not simultaneously zero;
(iii) No consumer is satiated;
(iv) There are no externalities in either production or consumption; that is, the output levels of each producer are independent of the output levels of the other producers, and the utility of each consumer is independent of the consumption choices of the others.

4.3.1 Pareto optimality for production

In the interest of simplicity, let us consider a situation in which there are only two producers ($m=2$) using two resources ($r=2$), each producing one of the two goods ($n=2$). Then we do not need to use (4.3.1), but, instead, write

$$q_1=f_1(x_{11},x_{12}), \qquad q_2=f_2(x_{21},x_{22}),$$

where $x_{11}+x_{21}=\bar{x}_1$ and $x_{12}+x_{22}=\bar{x}_2$ are the given resources. With a given level of output q_2^o, we wish to maximize q_1. The problem is none other than one of maximizing a single function with constraint. Hence we form the Lagrangian

$$L=f_1(x_{11},x_{12})+\lambda\left[f_2(\bar{x}_1-x_{11},\bar{x}_2-x_{12})-q_2^o\right].$$

As in Chapter 3, when we found the necessary condition, we equate its partial derivatives with respect to its three arguments x_{11}, x_{12}, and λ, and obtain the result

$$\frac{\partial f_1/\partial x_{11}}{\partial f_1/\partial x_{12}}=\frac{\partial f_2/\partial x_{21}}{\partial f_2/\partial x_{22}}, \qquad (4.3.5)$$

which is exactly (4.2.3) of the last section.

4.3.2 Pareto optimality for consumption

Similarly, we consider the simple situation of two consumers ($c=2$) and two goods ($n=2$) which are consumed by them, with $q_{11}+q_{21}=\bar{q}_1$ and $q_{12}+q_{22}=\bar{q}_2$ fixed. Given the level of satisfaction u_2^o of the second consumer, we try to maximize u_1 by the same procedure as above. Since we ignore primary factors, we do not need (4.3.2) but, instead, write

$$u_1=u_1(q_{11},q_{12})$$

and

$$u_2=u_2(q_{21},q_{22})=u_2(\bar{q}_1-q_{11},\bar{q}_2-q_{12})=u_2^o.$$

The Lagrangian is

$$L=u_1(q_{11},q_{12})+\lambda\left[u_2(\bar{q}_1-q_{11},\bar{q}_2-q_{12})-u_2^o\right].$$

[3] Actually, we only need to assume that these functions have continuous first derivatives; second derivatives are needed if we consider sufficient conditions.

After equating the derivatives to zero, we get

$$\frac{\partial u_1/\partial q_{11}}{\partial u_1/\partial q_{12}} = \frac{\partial u_2/\partial q_{21}}{\partial u_2/\partial q_{22}}, \tag{4.3.6}$$

which is (4.2.4).

4.3.3 Pareto optimality in the economy

Let our economy be composed of c consumers, m producers, n goods, and r resources, as stated at the beginning of this section. In addition to the assumption we made earlier, we assume that each consumer consumes all goods and each producer uses all given resources to produce all goods. We wish to maximize the utility of one consumer, say the first consumer, while the utility levels of all other consumers are given at u_i^o, for $i=2,\dots,c$. In view of the constraints given by (4.3.1), (4.3.3), and (4.3.4), the Lagrangian is

$$L = u_1\left(q_{11}^*, \dots, q_{1n}^*, \bar{x}_{11} - x_{11}^*, \dots, \bar{x}_{1r} - x_{1r}^*\right)$$

$$+ \sum_{i=2}^{c} \lambda_i\left[u_i\left(q_{i1}^*, \dots, q_{in}^*, \bar{x}_{i1} - x_{i1}^*, \dots, \bar{x}_{ir} - x_{ir}^*\right) - u_i^o \right]$$

$$+ \sum_{h=1}^{m} \mu_h F_h\left(q_{h1}, \dots, q_{hn}, x_{h1}, \dots, x_{hr}\right)$$

$$+ \sum_{j=1}^{r} \alpha_j\left(\sum_{i=1}^{c} x_{ij}^* - \sum_{h=1}^{m} x_{hj}\right) + \sum_{k=1}^{n} \beta_k\left(\sum_{h=1}^{m} q_{hk} - \sum_{i=1}^{c} q_{ik}^*\right),$$

where λ_i, μ_h, α_j, and β_k are Lagrange multipliers. We need to find the following partial derivatives and equate them to 0:

$$\frac{\partial L}{\partial q_{1k}^*} = \frac{\partial u_1}{\partial q_{1k}^*} - \beta_k = 0, \qquad k=1,\dots,n; \tag{4.3.7}$$

$$\frac{\partial L}{\partial x_{1j}^*} = -\frac{\partial u_1}{\partial\left(\bar{x}_{1j} - x_{1j}^*\right)} + \alpha_j = 0, \qquad j=1,\dots,r; \tag{4.3.8}$$

$$\frac{\partial L}{\partial q_{ik}^*} = \lambda_i \frac{\partial u_i}{\partial q_{ik}^*} - \beta_k = 0, \qquad i=2,\dots,c; \quad k=1,\dots,n; \tag{4.3.9}$$

$$\frac{\partial L}{\partial x_{ij}^*} = -\lambda_i \frac{\partial u_i}{\partial\left(\bar{x}_{ij} - x_{ij}^*\right)} + \alpha_j = 0, \qquad i=2,\dots,c; \quad j=1,\dots,r; \tag{4.3.10}$$

$$\frac{\partial L}{\partial q_{hk}} = \mu_h \frac{\partial F_h}{\partial q_{hk}} + \beta_k = 0, \qquad h=1,\dots,m; \quad k=1,\dots,n; \tag{4.3.11}$$

$$\frac{\partial L}{\partial x_{hj}} = \mu_h \frac{\partial F_h}{\partial x_{hj}} - \alpha_j = 0, \qquad h=1,\dots,m; \quad j=1,\dots,r. \tag{4.3.12}$$

By taking the ratios of proper pairs of equations, we arrive at various conclusions. From (4.3.7), (4.3.9), and (4.3.11), we solve for β_k/β_l ($k,l=1,\ldots,n$):

$$\frac{\beta_k}{\beta_l} = \frac{\partial u_1/\partial q_{1k}^*}{\partial u_1/\partial q_{1l}^*} = \cdots = \frac{\partial u_c/\partial q_{ck}^*}{\partial u_c/\partial q_{cl}^*} = \frac{\partial F_1/\partial q_{1k}}{\partial F_1/\partial q_{1l}} = \cdots = \frac{\partial F_m/\partial q_{mk}}{\partial F_m/\partial q_{ml}}. \quad (4.3.13)$$

This states that the MRCS for all the consumers and the marginal rates of product transformation for all producers are equal for every pair of commodities.

From (4.3.8), (4.3.10), and (4.3.12), we solve for α_j/α_s ($j,s=1,\ldots,r$):

$$\frac{\alpha_j}{\alpha_s} = \frac{\partial u_1/\partial\left(\bar{x}_{1j}-x_{1j}^*\right)}{\partial u_1/\partial\left(\bar{x}_{1s}-x_{1s}^*\right)}$$

$$\vdots$$

$$= \frac{\partial u_c/\partial\left(\bar{x}_{cj}-x_{cj}^*\right)}{\partial u_c/\partial\left(\bar{x}_{cs}-x_{cs}^*\right)} \quad (4.3.14)$$

$$= \frac{\partial F_1/\partial x_{1j}}{\partial F_1/\partial x_{1s}} = \cdots = \frac{\partial F_m/\partial x_{mj}}{\partial F_m/\partial x_{ms}}.$$

This states that the MRCS for the consumers and the MRTS for the producers are equal for every pair of resources.

From the six equations given by (4.3.7)–(4.3.12) we can also solve for α_j/β_k ($j=1,\ldots,r$; $k=1,\ldots,n$):

$$\frac{\alpha_j}{\beta_k} = \frac{\partial u_1/\partial\left(\bar{x}_{1j}-x_{1j}^*\right)}{\partial u_1/\partial q_{1k}^*}$$

$$\vdots$$

$$= \frac{\partial u_c/\partial\left(\bar{x}_{cj}-x_{cj}^*\right)}{\partial u_c/\partial q_{ck}^*} \quad (4.3.15)$$

$$= \frac{\partial F_1/\partial x_{1j}}{\partial F_1/\partial q_{1k}} = \cdots = \frac{\partial F_m/\partial x_{mj}}{\partial F_m/\partial q_{mk}},$$

which states that the consumer's MRCS between resources and commodities are equal to the producers' rates of transforming the corresponding resources into commodities.

The Lagrange multipliers α_j and β_k are generally interpreted as efficiency prices. Thus any $\pi_j=a\alpha_j$ ($j=1,\ldots,r$) for resources and $p_k=a\beta_k$ ($k=1,\ldots,n$) for commodities, where $a>0$ is some proportionality factor, are efficiency prices.

4.4 Pareto optimality in general

In recent years there has been a certain amount of interest on the part of mathematicians in Pareto optimality and related topics. They have greatly generalized the analysis which we presented in the last section.

S. Smale in a series of papers extended the notion of Pareto optimum to a larger set of critical points of several differentiable functions to study the optimization of several functions. In Section 4.3 we assumed that each consumer consumed each commodity. Smale relaxed this assumption and further extended the analysis of the stability of price equilibria made by Debreu[4] and others. These papers as well as various others by such authors as L. Cesari, M.B. Suryanarayana, Y.H. Wan, C.P. Simon and C. Titus, P.L. Yu, G. Leitmann, M. Zeleny, *etc.*, study various aspects of the general problem of optimizing several functions, some going into Pareto optimality in control theory. The mathematical level of these studies is beyond that assumed in this volume, so a bibliography of these titles is not given. However, C.P. Simon's paper (to be published), lists some of them.

We state without proof the following two theorems. The proofs may be found in Simon, Chapter 8.[5] In each we take, as before, an economy with c consumers and n goods under the same four assumptions of the last section. We shall be concerned only with consumption, and shall let q_{ik} be the amount of the kth commodity that is available and consumed by the ith consumer. Hence (4.3.4) reads

$$\sum_{i=1}^{c} q_{ik} = b_k > 0, \quad \text{for } k = 1, \ldots, n. \tag{4.4.1}$$

We further assume that $\partial u_i / \partial q_{ik} > 0$ whenever $q_{ik} = 0$. (Thus each consumer would like to consume at least some of that commodity if it is not available to him.) Associated with each consumer is an n-dimensional commodity vector $\mathbf{q}_i = (q_{i1}, \ldots, q_{in})^T$, $i = 1, \ldots, c$, and we shall call the set consisting of c such vectors a *commodity bundle*. A commodity bundle $\mathbf{q}$, then, may be regarded as a point in the cn-dimensional Euclidean space $\mathbf{R}^{cn}$, and we let Ω be the space of feasible commodity bundles:

$$\Omega = \left\{ \mathbf{q} = (\mathbf{q}_1, \ldots, \mathbf{q}_c) \in \mathbf{R}^{cn} \middle| \text{each } \mathbf{q}_i \geqslant 0 \text{ and } \sum_{i=1}^{c} \mathbf{q}_i = \mathbf{b} = (b_1, \ldots, b_n)^T \right\} \subset \mathbf{R}^{cn}.$$

Theorem 4.4.1.

(a) *Suppose a commodity bundle $\bar{\mathbf{q}} = (\bar{\mathbf{q}}_1, \ldots, \bar{\mathbf{q}}_c)$ is a local Pareto optimum (not necessarily interior) for the functions $u_1, \ldots, u_c$. Then there exist nonnegative numbers $\alpha_1, \ldots, \alpha_c$ not all zero and numbers $\gamma_1, \ldots, \gamma_n$ not all zero such*

[4]Debreu, G., 1970. Economics with a finite set of equilibria, *Econometrica*, **38**, pp. 387–92.
[5]See also Malinvaud (1972), pp. 76–96.

that

$$
\alpha_i \nabla u_i(\bar{\mathbf{q}}_i) = \begin{bmatrix} \alpha_i \dfrac{\partial u_i}{\partial q_{i1}}\bigg|_{\bar{q}_{i1}} \\ \vdots \\ \alpha_i \dfrac{\partial u_i}{\partial q_{in}}\bigg|_{\bar{q}_{in}} \end{bmatrix} \leqslant \begin{bmatrix} \gamma_1 \\ \vdots \\ \gamma_n \end{bmatrix} \qquad \textit{for } i = 1,\ldots,c, \qquad (4.4.2)
$$

and

$$
\alpha_i \frac{\partial u_i}{\partial q_{ik}}\bigg|_{\bar{q}_{ik}} = \gamma_k \quad \textit{whenever } \bar{q}_{ik} \neq 0.
$$

If none of the commodity vectors $\bar{\mathbf{q}}_i$ in the bundle $\bar{\mathbf{q}}$ is zero (the components are not all zero) and if $\partial u_i / \partial q_{ij} > 0$ for all i and j and for all commodities $\mathbf{q}_j$ satisfying our general assumptions (not just those which are locally Pareto optimal), then the $\alpha_1,\ldots,\alpha_c$ and $\gamma_1,\ldots,\gamma_n$ of (4.4.2) are all positive.

(b) *If $\bar{\mathbf{q}}$ is an interior local Pareto optimum as we assumed in Section 4.3, then:*

(i) *There exist positive numbers $\alpha_1,\ldots,\alpha_c$ and numbers $\gamma_1,\ldots,\gamma_n$ not all zero such that*

$$
\begin{bmatrix} \alpha_i \dfrac{\partial u_i}{\partial q_{i1}}\bigg|_{\bar{q}_{i1}} \\ \vdots \\ \alpha_i \dfrac{\partial u_i}{\partial q_{in}}\bigg|_{\bar{q}_{in}} \end{bmatrix} = \begin{bmatrix} \gamma_1 \\ \vdots \\ \gamma_n \end{bmatrix} \qquad \textit{for } i = 1,\ldots,c; \qquad (4.4.3)
$$

(ii) *At $\bar{\mathbf{q}}$, the marginal rate of commodity substitution MRCS of good j for good k ($j,k = 1,\ldots,n$) is the same for all consumers,*

$$
\frac{\partial u_1 / \partial q_{1j}}{\partial u_1 / \partial q_{1k}} = \ldots = \frac{\partial u_c / \partial q_{cj}}{\partial u_c / \partial q_{ck}}, \qquad (4.4.4)
$$

where the derivatives are all evaluated at $\mathbf{q}_i = \bar{\mathbf{q}}_i$. $\qquad\qquad\square$

As the reader must recognize, this theorem strengthens our results of the last section and (4.4.4) is the first part of (4.4.13).

The next theorem gives a sufficient condition.

Theorem 4.4.2. *We have the same assumptions as in Theorem 4.4.1.*

(a) *Let $\bar{\mathbf{q}}$ be an interior point in Ω, the space of feasible commodity bundles. $\bar{\mathbf{q}}$ is a local Pareto optimum if and only if (i) there exist positive numbers $\alpha_1,\ldots,\alpha_c$ and numbers $\lambda_1,\ldots,\lambda_n$ not all zero such that (4.4.3) holds* *and* (ii) $\bar{\mathbf{q}}$ *maximizes some* u_i *over a subspace* S *of* Ω, *where*

$$S = \{\mathbf{q} \in \Omega | u_l(\mathbf{q}) = u_l(\bar{\mathbf{q}}), \, l = 1, \dots, c, l \neq i\}.$$

(b) *If each* u_i, $i = 1, \dots, c$ *is concave and there exist positive* $\alpha_1, \dots, \alpha_c$ *such that* (4.4.2) *holds, then* $\bar{\mathbf{q}}$ *is a Pareto optimum for the utility functions* $u_1, \dots, u_c$. □

Exercises

4.1 Consider a two-person, two-commodity pure exchange (trade) economy in which each person accepts prices as given. Person A has an initial endowment of 50 units of good 1 and 100 units of good 2. Person B has 90 units of 1 and 60 units of 2. The utility functions of the two are $u_A = q_{A1}^2 q_{A2}$ and $u_B = q_{B1} q_{B2} = (140 - q_{A1})(160 - q_{A2})$.

(a) Draw an Edgeworth–Bowley diagram to show that in general there exists some trade between A and B so that both may obtain higher levels of utility.

(b) Derive the equation of the contract curve as a function of q_{A1} and q_{A2}.

(c) In competitive equilibrium,

$$\frac{\partial u_i / \partial q_{i1}}{\partial u_i / \partial q_{i2}} = \frac{p_1}{p_2}$$

for $i = A, B$, where p_1 and p_2 are the prices of goods 1 and 2 respectively. Let $p_2 = 1$. Under what conditions will trade take place?

4.2 In the economy of the above problem, let the initial endowments of goods 1 and 2 be $\bar{q}_1$ and $\bar{q}_2$ respectively, with $q_{A1} + q_{B1} = \bar{q}_1$ and $q_{A2} + q_{B2} = \bar{q}_2$, and let the utility functions be $u_A = q_{A1}^\alpha q_{A2}$ and $u_B = q_{B1}^\beta q_{B2}$. Determine the condition on α and β so that the contract curve will be a straight line.

4.3 Fill in the details to verify (4.3.5), (4.3.6), (4.3.13), (4.3.14), and (4.3.15).

4.4 In a two-person, two-commodity economy, assume now that *externality* exists, that is, the utility of one consumer depends upon the consumption of the other. Let $q_{A1} + q_{B1} = \bar{q}_1$ and $q_{A2} + q_{B2} = \bar{q}_2$, and let the utilities of the two persons be $u_A = u_A(q_{A1}, q_{A2}, q_{B1}, q_{B2})$ and $u_B = u_B(q_{A1}, q_{A2}, q_{B1}, q_{B2})$. Find the necessary condition for Pareto optimality.

4.5 Let our economy consist of two consumers A and B, one producer, one primary factor X, one ordinary good Q_1, and one public good Q_2. The public good is consumed by A and B jointly; no one's utility is diminished by the other person's consumption of it and both get satisfaction from q_2, the total output of Q_2. Let q_{i1} ($i = A, B$) be the consumption of Q_1 by the ith consumer, with $q_{A1} + q_{B1} = q_1$, let $\bar{x}_i$ ($i = A, B$) be the ith consumer's endowment of X, and let x_i be the amount that he supplies for production, with $x_A + x_B = x$. The production function is $F(q_1, q_2, x) = 0$ and the utility functions are

$$u_i = u(q_{i1}, q_2, \bar{x}_i - x_i), \qquad i = A, B.$$

Show that the necessary conditions for Pareto optimality imply that:

(1) The sum of the MRCS's of Q_1 for Q_2 for the consumers equals the MRPT (marginal rate of product transformation) of Q_1 for Q_2 in production,

that is,

$$\frac{\partial u_A / \partial q_2}{\partial u_A / \partial q_{A1}} + \frac{\partial u_B / \partial q_2}{\partial u_B / \partial q_{B1}} = \frac{\partial F / \partial q_2}{\partial F / \partial q_1};$$

(2) The MRCS of X for Q_1 for each consumer equals the negative of the marginal product MP of X in the production of Q_1, that is,

$$\frac{\partial u_A / \partial (\bar{x}_A - x_A)}{\partial u_A / \partial q_{A1}} = \frac{\partial u_B / \partial (\bar{x}_B - x_B)}{\partial u_B / \partial q_{B1}} = -\frac{\partial F / \partial x}{\partial F / \partial q_1};$$

(3) The sum of the MRCS's of X for Q_2 equals the reciprocal of the MP of X in the production of Q_2, that is,

$$\frac{\partial u_A / \partial q_2}{\partial u_A / \partial (\bar{x}_A - x_A)} + \frac{\partial u_B / \partial q_2}{\partial u_B / \partial (\bar{x}_B - x_B)} = -\frac{\partial F / \partial q_2}{\partial F / \partial x}.$$

5 Linear Programming

5.1 General nature of linear programming problems

One main feature of the classical approach to the problems of optimization is the assumption of a continuous derivative which equals zero at an interior point of maximum or minimum. This is the so-called first-order condition of marginal analysis in economics, for instance. In the case of maximization under constraint, the constraints are precise and stated in the form of equations, and the number of constraint equations is less than or equal to the number of variables under consideration. As we showed in Section 3.2 for the case of two variables, we attempt to find the point of tangency of the curves representing the function to be maximized and the constraint equation. This again requires the existence of the derivative. Calculus techniques fail if the graph of the function is discontinuous or kinked (has undefined slope) or if the constraints are stated not as equations but as inequalities or if the number of constraints is greater than the number of variables. Furthermore, we considered local, instead of global, extrema only.

We shall now consider linear programming problems, and in these (i) the optimum obtained is a global one, (ii) we are not concerned with the derivative of any function, (iii) the constraints are not necessarily stated as equations, and (iv) the number of constraints may exceed the number of variables.

Broadly speaking, programming consists of determining those values of the variables $x_1, x_2, \ldots, x_n$ subject to certain constraints on these values, which will maximize (minimize) a function called the *objective function*. Used in this sense, the problem of classical maximization as discussed in the previous chapters is also a mathematical programming problem. However, popular usage of the word programming usually refers to the process of finding an optimal (maximal or minimal) value of a given objective function $f(x_1, x_2, \ldots, x_n)$ of n variables (usually, but not necessarily required to be nonnegative) $x_1, x_2, \ldots, x_n$ ($x_j \geqslant 0$ for $j = 1, 2, \ldots, n$) subject to m inequality constraints of the type $g^i(x_1, x_2, \ldots, x_n) \leqslant b_i$, $i = 1, 2, \ldots, m$, where the b_i are constants, and m does not have to be less than or equal to n. If the objective function f and the constraint functions g^i are all linear, that is, if they are of the form

$$f(\mathbf{x}) = f(x_1, x_2, \ldots, x_n) = c_1 x_1 + c_2 x_2 + \ldots + c_n x_n + d,$$

and   (5.1.1)

$$g^1(\mathbf{x}) = a_{11} x_1 + a_{12} x_2 + \ldots + a_{1n} x_n \leqslant b_1,$$
$$g^2(\mathbf{x}) = a_{21} x_1 + a_{22} x_2 + \ldots + a_{2n} x_n \leqslant b_2,$$
$$\vdots$$
$$g^m(\mathbf{x}) = a_{m1} x_1 + a_{m2} x_2 + \ldots + a_{mn} x_n \leqslant b_m,$$

then we call the problem a *linear programming problem*. In (5.1.1), we use the notation $\mathbf{x}$ to represent the n-dimensional vector $\mathbf{x} = (x_1, x_2, \ldots, x_n)^{\mathrm{T}}$.

Although the assumption of linearity is a stringent one, there is a surprisingly large class of practical problems which lend themselves to solution by the techniques of linear programming, and many nonlinear problems may be successfully approximated by linear programs.

In general, a typical linear programming problem can be stated in the form: Minimize

$$z_y = b_1 y_1 + b_2 y_2 + \ldots + b_m y_m + d$$

subject to   (5.1.2)

$$a_{11} y_1 + a_{21} y_2 + \ldots + a_{m1} y_m \geqslant c_1,$$
$$\vdots$$
$$a_{1n} y_1 + a_{2n} y_2 + \ldots + a_{mn} y_m \geqslant c_n,$$

and

$$y_i \geqslant 0, \qquad i = 1, 2, \ldots, m.$$

Or, as is often the case, one might wish to maximize the objective function

$$z_x = c_1 x_1 + c_2 x_2 + \ldots + c_n x_n + d$$

subject to   (5.1.3)

$$a_{11} x_1 + a_{12} x_2 + \ldots + a_{1n} x_n \leqslant b_1,$$
$$\vdots$$
$$a_{m1} x_1 + x_{m2} x_2 + \ldots + a_{mn} x_n \leqslant b_m,$$

and

$$x_j \geqslant 0, \qquad j = 1, 2, \ldots, n,$$

where the a_{ij}, b_i, c_j, and d are given constants.

We shall assume that the system of constraint inequalities is linearly independent, that is, no left-hand side of any one of the inequalities may be expressed as a linear combination of the left-hand sides of the remaining inequalities. This will in general be the case under consideration. We will look at dependency in a later section.

The function to be maximized (minimized) is called the *objective function* and the variables x_j to be solved are called *decision* or *activity variables*. Any solution consisting of those values of the x_j which satisfy the nonnegative conditions and the inequality constraints is called a *feasible solution*. Any feasible solution which also maximizes (minimizes) the objective function is called an *optimal feasible solution*.

It is not necessary for the problem to assume exactly one or the other of the two forms above. The following techniques are frequently used to convert a stated problem to a more convenient form.

1. *Changing the sense of an inequality*.

$$\sum_{j=1}^{n} a_{ij}x_j \geqslant b_i$$

can be written as

$$\sum_{j=1}^{n} (-a_{ij})x_j \leqslant -b_i$$

and *vice versa*. For example,

$$3x_1 - 4x_2 \geqslant -5$$

is equivalent to

$$-3x_1 + 4x_2 \leqslant 5.$$

2. *Converting an equation to inequalities*.

Suppose a requirement is stated as an equation, such as

$$2x_1 + 3x_2 - 4x_3 = 5.$$

This is equivalent to

$$2x_1 + 3x_2 - 4x_3 \leqslant 5 \quad \text{and} \quad 2x_1 + 3x_2 - 4x_3 \geqslant 5,$$

that is,

$$2x_1 + 3x_2 - 4x_3 \leqslant 5 \quad \text{and} \quad -2x_1 - 3x_2 + 4x_3 \leqslant -5.$$

To see this, notice, for example, that $x = 4$ is equivalent to the combined statements $x \leqslant 4$ and $x \geqslant 4$.

In general,

$$\sum_{j=1}^{n} a_{ij}x_j = b_i$$

can be written as

$$\sum_{j=1}^{n} a_{ij}x_j \leqslant b_i \quad \text{and} \quad \sum_{j=1}^{n} \alpha_j x_j \leqslant \beta$$

where

$$\alpha_j = - \sum_{i=1}^{m} a_{ij} \quad \text{and} \quad \beta = - \sum_{i=1}^{m} b_i$$

for $i = 1, 2, \ldots, m$ and $j = 1, 2, \ldots, n$.

For example, the system

$$x_1 - 2x_2 \qquad = -1,$$
$$3x_1 \qquad + x_3 = 6,$$
$$3x_2 - 2x_3 = 1,$$

is equivalent to the system

$$x_1 - 2x_2 \qquad \leqslant -1,$$
$$3x_1 \qquad + x_3 \leqslant 6,$$
$$3x_2 - 2x_3 \leqslant 1,$$
$$-4x_1 - x_2 + x_3 \leqslant -6.$$

We see from the above examples that any system of linear equalities or inequalities can be represented as a system of linear inequalities of the same sense or direction.

3. *Converting an inequality to an equation.*

$$3x_1 + 2x_2 - 4x_3 \leqslant 2$$

is equivalent to

$$3x_1 + 2x_2 - 4x_3 + x_4 = 2, \quad \text{where } x_4 \geqslant 0,$$

and

$$7x_1 + x_2 + 3x_3 - 5x_4 \geqslant 6$$

is equivalent to

$$7x_1 + x_2 + 3x_3 - 5x_4 - x_5 = 6, \quad \text{where } x_5 \geqslant 0.$$

In general,

$$\sum_{j=1}^{n} a_{ij} x_j \leqslant b_i$$

may be written as

$$\sum_{j=1}^{n} a_{ij} x_j + y_i = b_i$$

and

$$\sum_{j=1}^{n} a_{ij} x_j \geqslant b_i$$

may be written as

$$\sum_{j=1}^{n} a_{ij} x_j - z_i = b_i,$$

where $y_i \geqslant 0$ and $z_i \geqslant 0$. We usually call y_i a *slack variable* and z_i a *surplus variable*.

4. *Changing the restrictions on the decision variables.*

Suppose that for some j, x_j is unrestricted (or free) in sign. We can write x_j as

$$x_j = x_j' - x_j'', \quad \text{where } x_j' \geqslant 0 \text{ and } x_j'' \geqslant 0$$

and let

$$x_j' = x_j \quad \text{and} \quad x_j'' = 0 \quad \text{when } x_j > 0$$

and

$$x_j'' = x_j \quad \text{and} \quad x_j' = 0 \quad \text{when } x_j < 0.$$

When we apply the simplex method, to be presented in Section 5.3, if there are several free variables we may either apply the same technique to each variable or we may use a single nonnegative variable y, the same for each free variable x_j, by writing

$$x_j = x_j' - y, \quad \text{where } x_j' \geqslant 0 \text{ and } y \geqslant 0,$$

as before.

For example: Maximize

$$z = 2x_1 + 5x_2$$

subject to

$$3x_1 + x_2 \leqslant 4$$

and

$$-x_1 + 5x_2 \leqslant 6,$$
$$x_1 \text{ and } x_2 \text{ free.}$$

This system may be expressed as: Maximize

$$z = 2x_1' - 2x_1'' + 5x_2' - 5x_2''$$

subject to

$$3x_1' - 3x_1'' + x_2' - x_2'' \leqslant 4,$$
$$-x_1' + x_1'' + 5x_2' - 5x_2'' \leqslant 6,$$

$$x_1' \geqslant 0, \quad x_1'' \geqslant 0, \quad x_2' \geqslant 0, \quad x_2'' \geqslant 0, \quad x_1'x_1'' = 0, \quad x_2'x_2'' = 0;$$

or as

$$z = 2x_1' + 5x_2' - 7y$$

subject to

$$3x_1' + x_2' - 4y \leqslant 4,$$
$$-x_1' + 5x_2' - 4y \leqslant 6,$$

$$x_1' \geqslant 0, \quad x_2' \geqslant 0, \quad y \geqslant 0, \quad x_1'y = 0, \quad x_2'y = 0.$$

Suppose the lower bound for a decision variable x_j is not zero but $d_j \neq 0$, that is,

$$x_j \geqslant d_j.$$

This can be taken care of by introducing a new variable

$$y_j = x_j - d_j.$$

Then $y_j \geqslant 0$.

More will be said in a later section on methods to deal with free or lower bounded variables.

It will suffice to conclude from the above transformations that it is generally possible to state a linear programming problem in the form (5.1.2) or (5.1.3). Either of these may be solved fairly easily by the simplex method which will be presented in Section 5.3. Actually, in employing the simplex method, surplus or slack variables are introduced into (5.1.2) or (5.1.3) so that the requirements read:

$$a_{11}x_1 + a_{12}x_2 + \ldots + a_{1n}x_n + a_{1n+1}x_{n+1} + \ldots + a_{1n+m}x_{n+m} = b_1$$

$$a_{21}x_1 + a_{22}x_2 + \ldots + a_{2n}x_n + a_{2n+1}x_{n+1} + \ldots + a_{2n+m}x_{n+m} = b_2$$

$$\vdots \tag{5.1.4}$$

$$a_{m1}x_1 + a_{m2}x_2 + \ldots + a_{mn}x_n + a_{mn+1}x_{n+1} + \ldots + a_{mn+m}x_{n+m} = b_m$$

$$x_j \geqslant 0, \quad j = 1, 2, \ldots, m+n.$$

5.2 Some geometric and algebraic interpretations

Let us consider a simple example involving only two decision variables: Maximize

(0) $$f(x_1, x_2) = x_1 + 4x_2$$

subject to

(1) $$2x_1 + 3x_2 \leqslant 19,$$

(2) $$-3x_1 + 2x_2 \leqslant 4, \tag{5.2.1}$$

(3) $$x_1 + x_2 \leqslant 8,$$

$$x_1 \geqslant 0, \quad x_2 \geqslant 0.$$

The graph of the problem is presented in Figure 5.2.1. The straight lines (1), (2), and (3), and the positive x_1- and x_2-axes form the boundaries of the closed and bounded region R determined by the constraint inequalities. The coordinates of every point within this region satisfy the requirements and thus form a feasible solution. R is actually a convex set and is the set of all feasible solutions of our problem. More will be said on convex sets shortly.

The four parallel lines (dashed) in the figure show four different values of the objective function:

$$f(x_1, x_2) = x_1 + 4x_2 = 8 \qquad (x_1 = 0, \ x_2 = 2);$$

$$f(x_1, x_2) = 16 \qquad (x_1 = 8, \ x_2 = 2);$$

$$f(x_1, x_2) = 22 \qquad (x_1 = 2, \ x_2 = 5);$$

and $\qquad f(x_1, x_2) = 36 \qquad (x_1 = 20, \ x_2 = 4).$

The function $f(x_1, x_2) = x_1 + 4x_2$ defines a family of parallel lines all with slope $-\frac{1}{4}$. For our problem, we consider only those points in the feasible region R which intercept this family of lines, and we see that the maximum value of the function which satisfies the constraint requirements is reached at the point (2,5), which is a corner or extreme point of the region R.

This example illustrates the situation which generally prevails in a linear programming problem. The function to be maximized (minimized) is a linear function whose graph is a family of parallel straight lines in two-space or hyperplanes in n-space. Some of these hyperplanes intersect the convex set of feasible solutions. From these we select the one which gives the maximum (minimum) value for the function. As we saw in the previous example, the point of intersection of the convex set and the linear function is a corner or extreme point of the convex set. This will be proved to be the case in general.

Lest the above example lead one into thinking that a linear function will always assume maximum (minimum) values at the extreme points of the convex set of feasible solutions, here is an example of a function which has neither a maximum nor a minimum:

Maximize (or minimize)

$$z = f(x_1, x_2) = 2x_1 + x_2$$

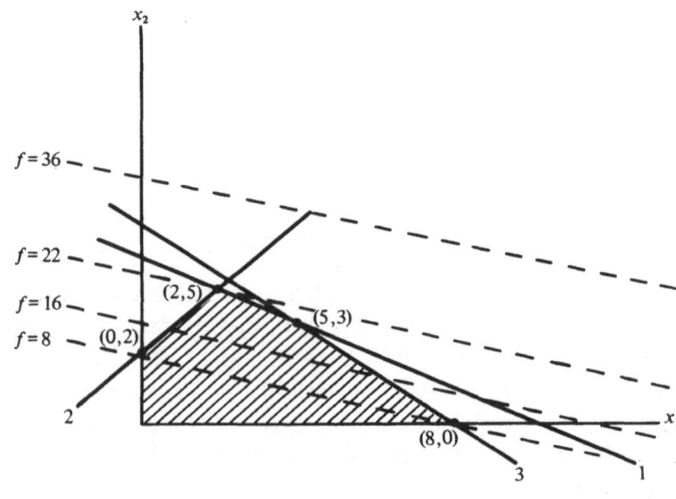

Figure 5.2.1.

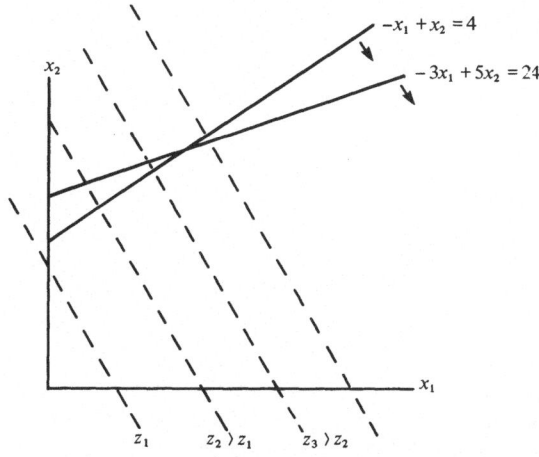

Figure 5.2.2.

subject to

$$-x_1 + x_2 \leqslant 4,$$

$$-3x_1 + 5x_2 \leqslant 24,$$

$$x_1 > 0, \qquad x_2 > 0.$$

The above convex set of solutions is neither bounded nor closed and the function has no finite maximum or minimum. If the restraints were changed to $x_1 \geqslant 0$ and $x_2 \geqslant 0$, then the function would have a minimum at the origin.

It will be shown later in this section that if the function has a finite maximum (minimum), then at least one such optimal solution is to be found at a corner (extreme) point of the convex set of feasible solutions. This being the case, the task of finding a feasible optimum for the objective function is reduced to examining the extreme points of the convex set. Since there are m constraints, a finite number, which determine the convex set, the number of extreme points will also be finite. The most common method used in solving such problems, to be presented in the next section, is the simplex method, which is essentially a systematic procedure for testing the extreme points.

Before we look at the simplex method, we will discuss some of the rationale behind certain aspects of the technique.

First, as we mentioned at the end of the previous section, if the constraints are stated as inequalities, as would frequently be the case, the usual practice is to introduce slack variables and to restate them in the form of equations.[1] Thus, for the example associated with Figure 5.2.1, the

---

[1] Even if an original constraint is stated as an equation, an artificial variable may be added to aid in the solution.

problem becomes:
Maximize

$$f(x_1, x_2, x_3, x_4, x_5) = x_1 + 4x_2 + 0x_3 + 0x_4 + 0x_5$$

subject to   (5.2.2)

$$2x_1 + 3x_2 + x_3 + 0x_4 + 0x_5 = 19,$$

$$-3x_1 + 2x_2 + 0x_3 + x_4 + 0x_5 = 4,$$

$$x_1 + x_2 + 0x_3 + 0x_4 + x_5 = 8,$$

$$x_j \geqslant 0, \quad j = 1, 2, 3, 4, 5.$$

A few words should be said on the dimensions of the solution space. The problem originally contained only two variables and was obviously in two-dimensional space. With the introduction of the three slack variables, it would appear that we now may have a different problem in five-dimensional space. Actually, (5.2.1) and (5.2.2) are algebraically equivalent, as will be shown in the following examples.

Consider the two systems:

$$0 \leqslant x_1 \leqslant 4,$$ (5.2.3)

and

$$x_1 + x_2 = 4,$$ (5.2.4)

$$x_1 \geqslant 0, \quad x_2 \geqslant 0.$$

The set of all ordered pairs (x_1, x_2) which satisfy (5.2.4) form the slanted line shown in Figure 5.2.3(a). It is not the same as the segment in Figure 5.2.3(b), which consists of all points that satisfy (5.2.3). However, consider the projection of the upper slanted line onto the x_1-axis, namely, all ordered pairs of the form $(x_1, 0)$, with $0 \leqslant x_1 \leqslant 4$. This is the darkened segment on the x_1-axis of Figure 5.2.3(a). We may make a one-to-one association between the points on the horizontal segment of Figure 5.2.3(a) and those on the segment of Figure 5.2.3(b), that is, we associate the pair

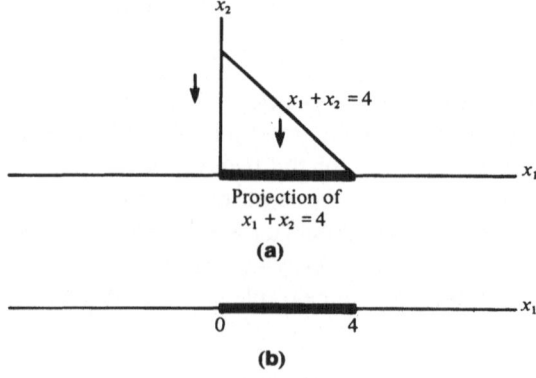

Figure 5.2.3

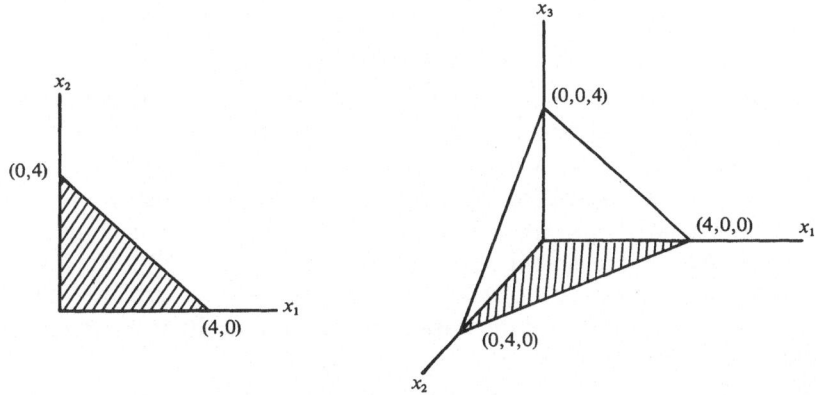

Figure 5.2.4

$(x_1, 0)$ in two-space with the point x_1 in one-space. With this interpretation, the set of x_1 values in the ordered pairs satisfying (5.2.4) is equivalent to the set of x_1 values satisfying (5.2.3).

Consider another example:

$$x_1 + x_2 \leqslant 4,$$
$$x_1 \geqslant 0, \qquad x_2 \geqslant 0,$$

(5.2.5)

and

$$x_1 + x_2 + x_3 = 4,$$
$$x_1 \geqslant 0, \qquad x_2 \geqslant 0, \qquad x_3 \geqslant 0.$$

(5.2.6)

The graph of (5.2.5) is the shaded triangle of Figure 5.2.4(a) and the graph of (5.2.6) is the triangular plane in the first octant in three-space of Figure 5.2.4(b). The projection of the plane onto the $x_1 x_2$-plane, namely, all triplets of the form $(x_1, x_2, 0)$, with $x_1 + x_2 \leqslant 4$ and $x_1, x_2 \geqslant 0$, is the shaded portion beneath it. Again, there is a one-to-one correspondence between the shaded region in two-space and that in three-space.

In general, a similar projection may be made from each hyperplane specified by an equation in $n + m$-space onto a half-space in n-space (specified by an inequality of the form $a_{i1} x_1 + a_{i2} x_2 + \ldots + a_{in} x_n \leqslant b_i$) by setting m of the variables equal to zero. A hyperplane in $n + m$-space is defined to be a linear space of dimension $n + m - 1$.[2] For example, the equation $a_1 x_1 + a_2 x_2 = b$ defines a line or linear space of dimension 1 $(= 2 - 1)$ in two-space. The intersection of two hyperplanes, each of dimension $n + m - 1$ in $n + m$-space, is a linear space of dimension $n + m - 2$, assuming that no equation is a linear combination of the others. For example, the intersection of the two hyperplanes in three-space defined by

$$x_1 + x_2 + x_3 = 4$$

[2] A set L is a linear space if, for any points $\mathbf{a}$ and $\mathbf{b} \in L$, $\mathbf{a} + \mathbf{b} \in L$, and if, for any real number k, $k\mathbf{a} \in L$.

and

$$4x_1 - 2x_2 + 3x_3 = 8$$

(each of dimension 2), is the line $6x_1 + 5x_3 = 16$, which is of dimension 1. Thus, if there are m intersecting hyperplanes, each of dimension $n + m - 1$, their intersection is a linear space of dimension $n + m - m = n$.

The convex set specified by the system of inequalities of the form $\sum_{j=1}^{n} a_{ij} x_j \geq b_i$ or $\sum_{j=1}^{n} a_{ij} x_j \leq b_i$ in (5.1.2) or (5.1.3) is of dimension n. When these inequalities are stated as equations of the form $\sum_{j=1}^{n+m} a_{ij} x_j = b_i$ in (5.1.4) by the introduction of m slack variables, each hyperplane specified by each such equation is of dimension $n + m - 1$, and the intersection of m of them is of dimension n.

The reader is cautioned not to confuse the dimension of the solution space with the dimension of the constraints or requirements space. Consider again (5.2.2), in which the revised solution space is of dimension 5, but the requirements space is of dimension 3, as can be seen more easily if the constraint equations are written in vector form as follows:

$$\begin{bmatrix} 2 \\ -3 \\ 1 \end{bmatrix} x_1 + \begin{bmatrix} 3 \\ 2 \\ 1 \end{bmatrix} x_2 + \begin{bmatrix} 1 \\ 0 \\ 0 \end{bmatrix} x_3 + \begin{bmatrix} 0 \\ 1 \\ 0 \end{bmatrix} x_4 + \begin{bmatrix} 0 \\ 0 \\ 1 \end{bmatrix} x_5 = \begin{bmatrix} 19 \\ 4 \\ 8 \end{bmatrix}. \qquad (5.2.7)$$

In general, the system of equations

$$\sum_{j=1}^{n+m} a_{ij} x_j = b_i, \qquad i = 1, 2, \ldots, m$$

may be expressed in vector form as

$$\begin{bmatrix} a_{11} \\ a_{21} \\ \vdots \\ a_{m1} \end{bmatrix} x_1 + \begin{bmatrix} a_{12} \\ a_{22} \\ \vdots \\ a_{m2} \end{bmatrix} x_2 + \ldots + \begin{bmatrix} a_{1n+m} \\ a_{2n+m} \\ \vdots \\ a_{mn+n} \end{bmatrix} x_{n+m} = \begin{bmatrix} b_1 \\ b_2 \\ \vdots \\ b_m \end{bmatrix}. \qquad (5.2.8)$$

Each x_j is a coefficient of the vector equation

$$\mathbf{v}_1 x_1 + \mathbf{v}_2 x_2 + \ldots + \mathbf{v}_{n+m} x_{n+m} = \mathbf{b},$$

where

$$\mathbf{v}_j = \begin{bmatrix} a_{1j} \\ a_{2j} \\ \vdots \\ a_{mj} \end{bmatrix}, \qquad j = 1, 2, \ldots, n+m$$

and

$$\mathbf{b} = \begin{bmatrix} b_1 \\ b_2 \\ \vdots \\ b_m \end{bmatrix}$$

are m-dimensional vectors.

In other words, the nonnegative numbers $x_1, x_2, \ldots, x_{n+m}$ form the coordinates of a point $(x_1, x_2, \ldots, x_{n+m})^{\mathsf{T}}$ in the feasible solution set in the $n+m$-dimensional solution space if and only if they are the coefficients of the vector equation (5.2.8), that is, if and only if the vector $\mathbf{b}$ may be expressed as a nonnegative linear combination of the vectors $\mathbf{v}_j$. More specifically, $\mathbf{b}$ is to be expressed as a nonnegative linear combination of a set of independent vectors in the requirements space. If the requirements space is of dimension m, then no set containing more than m vectors may be linearly independent, so that at most m of the x_j may be positive, while the other n must be zero.

Let us next look at a simple problem involving three variables in which slack variables will be introduced to change the constraining inequalities into equations. Maximize

$$z = x_1 + 3x_2 + 4x_3$$

subject to   (5.2.9)

$$3x_1 + 2x_2 \qquad \leqslant 13,$$

$$x_2 + 3x_3 \leqslant 17,$$

$$2x_1 + x_2 + x_3 \leqslant 13,$$

$$x_j \geqslant 0, \qquad j = 1, 2, 3.$$

The feasible solutions are represented by points within or on the boundary of the solid $OABCDEFG$ in 3-space which is enclosed by the six planes (see Figure 5.2.5)

$$
\begin{array}{ll}
x_1 = 0 & (OBCD), \\
x_2 = 0 & (ODEFA), \\
x_3 = 0 & (OAB), \\
3x_1 + 2x_2 = 13 & (ABCGF), \\
x_2 + 3x_3 = 17 & (CDEG), \\
2x_1 + x_2 + x_3 = 13 & (EFG).
\end{array}
$$

From analytic geometry, we know that the perpendicular distance of the plane

$$\frac{x_1}{\sqrt{26}} + \frac{3x_2}{\sqrt{26}} + \frac{4x_3}{\sqrt{26}} = p$$

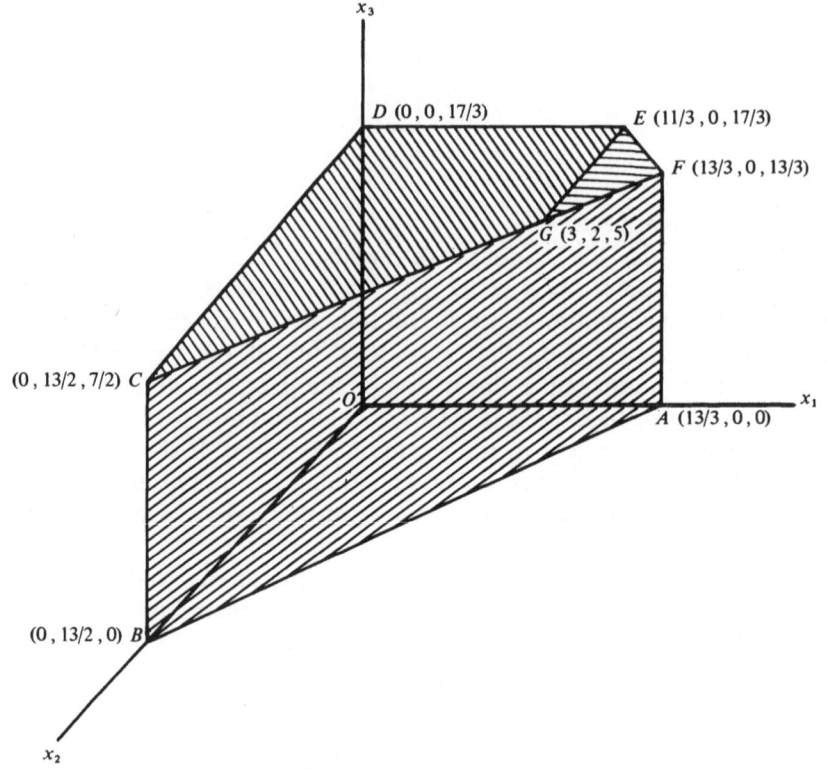

Figure 5.2.5.

from the origin is p. Since the function to be maximized, z, is equal to $p\sqrt{26}$, we can think of a family of parallel planes and finding that one which is within or on the boundary of the solid $OABCDEFG$ and is furthest from the origin. It is obvious that such a plane must pass through at least one corner of this solid. It either passes through exactly one corner, or it passes through an edge (or side) with the same value for z along this edge (or side) including its corner points. Therefore, our search for the maximum value of z may be confined to the corner points.

When we add slack variables, the inequalities in (5.2.9) become

$$
\begin{aligned}
3x_1 + 2x_2 \quad + x_4 \qquad\qquad &= 13, \\
x_2 + 3x_3 \quad + x_5 \quad &= 17, \\
2x_1 + x_2 + x_3 \qquad\quad + x_6 &= 13,
\end{aligned}
\tag{5.2.10}
$$

which may be alternately written as

$$
\begin{aligned}
x_4 &= 13 - 3x_1 - 2x_2, \\
x_5 &= 17 - x_2 - 3x_3, \\
x_6 &= 13 - 2x_1 - x_2 - x_3.
\end{aligned}
\tag{5.2.10a}
$$

An edge of the solid $OABCDEFG$ is obtained by setting two of the six variables equal to zero. For example, the edge

CD	is from	$x_1=0$	and	$x_5=0$,
AB	is from	$x_3=0$	and	$x_4=0$,
EF	is from	$x_2=0$	and	$x_6=0$,

and so on. A corner of the solid is obtained by setting three of the variables equal to zero. For example, the corner A, which is the intersection of AB ($x_3=0$, $x_4=0$) and AF ($x_2=0$, $x_4=0$), is obtained by setting $x_2=0$, $x_3=0$, and $x_4=0$. Similarly, we get the point G ($x_1=3$, $x_2=2$, $x_3=5$) by setting $x_4=0$, $x_5=0$, and $x_6=0$, since it is the intersection of the edges CG ($x_4=0$, $x_5=0$) and EG ($x_5=0$, $x_6=0$).

For the example illustrated in Figure 5.2.1 and (5.2.1) and (5.2.2), the corner points are obtained by setting any two of the five variables equal to zero. For example, the corner point $(0,2)$ is obtained from $x_1=0$ and $x_4=0$. When we substitute these zero values in for x_1 and x_4 and solve the system (5.2.2) we get the point $(0,2,13,0,6)$, which is projected onto the point $(0,2)$ in two-space. Of course, the coefficient of x_3, x_4, and x_5 in the objective function f is 0, so that these slack variables do not make any contribution to the value of the objective function.

The above geometric observations may be substantiated algebraically.

We have said previously that the set of all feasible solutions is a convex set. Let us introduce a few more definitions.

Definition 1. Convex set. Let $\mathbf{u}$ and $\mathbf{v}$ be n-dimensional vectors or points in n-space and k be any real number such that $0 \leqslant k \leqslant 1$. A set S is convex in that space if $k\mathbf{u}+(1-k)\mathbf{v} \in S$ for every k. Another way of saying the same thing is: A convex set S is a collection of points such that for any two points $\mathbf{u}=(u_1,u_2,\ldots,u_n)^\mathrm{T}$ and $\mathbf{v}=(v_1,v_2,\ldots,v_n)^\mathrm{T}$ in the collection, the line segment joining them is also in S. The line segment joining the two points is the collection of points $\mathbf{p}(k)=(p_1,p_2,\ldots,p_n)^\mathrm{T}$ whose coordinates are given by

$$p_i(k)=k\,u_i+(1-k)v_i \qquad i=1,2,\ldots,n, \qquad 0 \leqslant k \leqslant 1.$$

Definition 2. Extreme point (or corner, or vertex) of the convex set S. This is a point of S which does not lie on any segment joining two other points of S.

Definition 3. Basic solution. Let the system of requirements be stated as equations in the form given by (5.1.4). A basic solution is a solution of (5.1.4) obtained by setting n of the $n+m$ variables equal to zero and solving for the remaining m variables, assuming that the values of these m variables may be uniquely determined (that is, if the determinant of the coefficients of these m variables is not equal to zero).[3] From the preceding paragraphs, it is seen that a basic solution is an extreme point of the convex set.

Definition 4. Basic variables or basis. These are the m variables which are not set equal to zero. From the discussion about the dimension of the requirements space, these variables are the coefficients of the m independent vectors which form the basis for the m-dimensional requirements space. These basic variables are sometimes called *dependent variables*, while the nonbasic ones which are set equal to zero are called *independent variables* because each basic variable may be expressed in terms of the nonbasic ones, as shown in (5.2.10a), if we choose x_4, x_5, and x_6 as the initial basic variables.

Definition 5. Basic feasible solution. This is a basic solution of (5.1.4) which also satisfies the nonnegative requirements that $x_j \geqslant 0$. Notice that a basic solution is not necessarily feasible. For the example in Figure 5.2.5 and (5.2.10), if we set x_1, x_3, and $x_5 = 0$, we will get a basic solution $(0, 17, 0, -21, 0, -4)$ which is not feasible because x_4 and x_6 are negative. Similarly, a feasible solution is not necessarily basic, since every point in the convex set defined by the constraining inequalities or equations is a feasible solution.

Definition 6. Optimal basic solution. This is a basic feasible solution which optimizes (maximizes or minimizes) the objective function.

The set of points which satisfy a linear inequality is called a *half-space*. It is quite easily proved that a half-space is a convex set and that the intersection of any collection of convex sets is convex. The set S of all feasible solutions of a linear programming problem is the intersection of a finite number of half-spaces and is therefore a convex set. This notion may be stated formally as a theorem, using the formulation given by (5.1.4).

Theorem 5.2.1. *The set S of feasible solutions of* (5.1.4) *is a convex set.*

Proof. Let

$$\mathbf{x}^{(F1)} = \left(x_1^{(F1)}, x_2^{(F1)}, \ldots, x_{n+m}^{(F1)} \right)^{\mathrm{T}}$$

and

$$\mathbf{x}^{(F2)} = \left(x_1^{(F2)}, x_2^{(F2)}, \ldots, x_{n+m}^{(F2)} \right)^{\mathrm{T}}$$

be any two feasible solutions. We wish to show that all points on the

---

[3]There is a theorem in algebra (Cramer) which states that m linear equations $\sum_{j=1}^{n} a_{ij} x_j = b_i$, for $i = 1, 2, \ldots, m$, in m variables has a unique solution if the determinant of coefficients $|a_{ij}|$ is not equal to zero. See Section 5.4.3 for a discussion of degeneracy and dependency.

segment joining these two points also belong to S. Now

$$\sum_{j=1}^{n+m} a_{ij}x_j^{(F1)} = b_i, \qquad i=1,2,\ldots,m, \tag{5.2.11}$$

$$\sum_{j=1}^{n+m} a_{ij}x_j^{(F2)} = b_i, \qquad i=1,2,\ldots,m, \tag{5.2.12}$$

and

$$x_j^{(F1)} \geq 0, \qquad x_j^{(F2)} \geq 0, \qquad j=1,2,\ldots,n+m. \tag{5.2.13}$$

If we multiply (5.2.11) by k and (5.2.12) by $1-k$, where $0 \leq k \leq 1$, and add them, we get

$$\sum_{j=1}^{n+m} a_{ij}\,x_j(k) = kb_i + (1-k)b_i = b_i,$$

where $x_j(k) = kx_j^{(F1)} + (1-k)x_j^{(F2)}$, and $x_j(k) \geq 0$. Therefore $\mathbf{x}(k) = (x_1(k), x_2(k), \ldots, x_{n+m}(k))^{\mathrm{T}}$ is a feasible solution and corresponds to a point on the segment joining the two points which correspond to $\mathbf{x}^{(F1)}$ and $\mathbf{x}^{(F2)}$. Since $\mathbf{x}^{(F1)}$ and $\mathbf{x}^{(F2)}$ are *any* two feasible solutions, this means that all points on the segment joining any two feasible solutions are also feasible. $\qquad\square$

We have stated that a basic solution corresponds to an extreme point of the convex set and have illustrated that geometrically by Figure 5.2.1 and Figure 5.2.5. This may also be proved, by the following two theorems.

Theorem 5.2.2. *A basic feasible solution of (5.1.4) is an extreme point of the convex set S of Theorem 5.2.1.*

Proof. Let $\mathbf{x}^{(BF)} = (x_1^{(BF)}, x_2^{(BF)}, \ldots, x_m^{(BF)}, 0, 0, \ldots, 0)^{\mathrm{T}}$ be a basic feasible solution. For convenience, we relabeled the variables so that the first m of them constitute the basis and the remaining n are set equal to 0. From the definition of a basic solution, we know that the coordinates $x_j^{(BF)}$ are obtained from the solution of

$$\sum_{j=1}^{m} a_{ij}x_j = b_i, \qquad i=1,2,\ldots,m, \tag{5.2.14}$$

and that the values of the $x_j^{(BF)}$ for $j=1,2,\ldots,n+m$ are uniquely determined. Suppose $\mathbf{x}^{(BF)}$ is not an extreme point but lies on a segment joining two other feasible solutions $\mathbf{x}^{(F1)}$ and $\mathbf{x}^{(F2)}$. Then

$$x_j^{(BF)} = kx_j^{(F1)} + (1-k)x_j^{(F2)}, \qquad j=1,2,\ldots,m,$$

and

$$0 = kx_j^{(F1)} + (1-k)x_j^{(F2)}, \qquad j=m+1,m+2,\ldots,m+n,$$

for $0 < k < 1$. Since $k > 0$, $1-k > 0$ and $x_j^{(F1)} \geq 0$, $x_j^{(F2)} \geq 0$, the only way for the sum $kx_j^{(F1)} + (1-k)x_j^{(F2)}$ to be 0 would be for

$$x_j^{(F1)} = x_j^{(F2)} = 0, \qquad j = m+1, m+2, \ldots, m+n.$$

Therefore $\mathbf{x}^{(F1)}$ and $\mathbf{x}^{(F2)}$ are also basic feasible solutions and the coordinates of $\mathbf{x}^{(F1)}$ and $\mathbf{x}^{(F2)}$ are also uniquely determined by the solution of (5.2.14). This simply means that

$$x_j^{(BF)} = x_j^{(F1)} = x_j^{(F2)}, \qquad j = 1, 2, \ldots, m+n,$$

or that $\mathbf{x}^{(F1)}$ or $\mathbf{x}^{(F2)}$ is the same point as $\mathbf{x}^{(BF)}$. That is to say, $\mathbf{x}^{(BF)}$ does not lie on a segment joining some two other feasible points. □

This theorem also tells us that if the set of positive numbers $x_1^{(BF)}, x_2^{(BF)}, \ldots, x_m^{(BF)}$ is a solution of the vector equation

$$\mathbf{v}_1 x_1 + \mathbf{v}_2 x_2 + \ldots + \mathbf{v}_m x_m = \mathbf{b},$$

where

$$\mathbf{v}_j = \begin{bmatrix} a_{1j} \\ a_{2j} \\ \vdots \\ a_{mj} \end{bmatrix}, \qquad \mathbf{b} = \begin{bmatrix} b_1 \\ b_2 \\ \vdots \\ b_m \end{bmatrix}, \qquad x_j > 0, \qquad j = 1, 2, \ldots, m,$$

and the $\mathbf{v}_j$ are linearly independent, then

$$\mathbf{x}^{(BF)} = \left(x_1^{(BF)}, x_2^{(BF)}, \ldots, x_m^{(BF)}, 0, \ldots, 0 \right)^{\mathrm{T}},$$

where the last n coordinates are zero, is an extreme point of the convex set S.

Theorem 5.2.3. *If* $\mathbf{x} = (x_1, x_2, \ldots, x_m, 0, \ldots, 0)^{\mathrm{T}}$ *is an extreme point of* S, *then* $\mathbf{x}$ *is a basic feasible solution of* (5.1.4).

Proof. It will suffice to show that there are linearly independent vectors $\mathbf{v}_1, \ldots, \mathbf{v}_m$ such that

$$\mathbf{v}_1 x_1 + \ldots + \mathbf{v}_m x_m = \mathbf{b}, \qquad x_j > 0, \qquad j = 1, 2, \ldots, m. \qquad (5.2.15)$$

Suppose that $\mathbf{v}_1, \ldots, \mathbf{v}_m$ are linearly dependent. Then we can find numbers $y_1, \ldots, y_m$ not all 0 such that

$$\mathbf{v}_1 y_1 + \ldots + \mathbf{v}_m y_m = 0. \qquad (5.2.16)$$

Multiplying (5.2.16) by a constant $t > 0$ and adding and subtracting the result to (5.2.15) gives

$$\mathbf{v}_1 (x_1 + ty_1) + \ldots + \mathbf{v}_m (x_m + ty_m) = \mathbf{b},$$
$$\mathbf{v}_1 (x_1 - ty_1) + \ldots + \mathbf{v}_m (x_m - ty_m) = \mathbf{b}.$$

Moreover, let t be so small that $x_j \neq ty_j > 0$ for $j = 1, 2, \ldots, m$. This is always possible since each $x_j > 0$. Then

$$\mathbf{x}^{(1)} = (x_1 + ty_1, \ldots, x_m + ty_m, 0, \ldots, 0)^{\mathrm{T}}$$

and

$$\mathbf{x}^{(2)} = (x_1 - ty_1, \ldots, x_m - ty_m, 0, \ldots, 0)^{\mathrm{T}}$$

are two feasible solutions and points of S. But $\mathbf{x} = \frac{1}{2}\mathbf{x}^{(1)} + \frac{1}{2}\mathbf{x}^{(2)}$, which contradicts the hypothesis that x is an extreme point. Therefore the $\mathbf{v}_j$ must be linearly independent. □

The next two theorems establish the importance of basic feasible solutions. We state them without proof.[4]

Theorem 5.2.4. *If a feasible solution exists, then a basic feasible solution exists.*

Theorem 5.2.5. *If the finite objective function has a finite optimum (maximum or minimum), then at least one optimal solution is a basic feasible solution.*

The simplex method to be presented in the next section makes precise use of the above theorems. It is a systematic search for the optimum by examining the extreme points (or basic feasible solutions) of the convex set S. Since the number m of constraints is finite, there are at most

$$\binom{n+m}{m} = \frac{(n+m)!}{m!n!}$$

extreme points.[5]

5.3 The simplex method

The name *simplex* is from the topological concept of an n-simplex which, very loosely speaking, is the analogue of a convex polygon.[6] Fortunately, this method, which is due to Dantzig, is also relatively simple and lends itself readily to calculations by the electronic computer. We illustrate it by solving the problem given in the previous section. Maximize

(0) $z - x_1 - 3x_2 - 4x_3 - 0x_4 - 0x_5 - x_6 = 0$

---

[4]The proofs are rather long. The interested reader is referred to Garvin (1960), pp. 12–25, or Aoki (1971), pp. 41–44.

[5]Any introductory probability or statistics book will have a section explaining the meaning and derivation of the binomial coefficient symbol $\binom{n}{r}$.

[6]For a definition of a simplex, see, for instance: Eilenberg, S., and Steenrod, 1952, *Foundations of Algebraic Topology*, p. 54. Princeton University Press, Princeton.

subject to

(1) $\qquad 3x_1 + 2x_2 \quad\;\; + \underline{x}_4 \qquad\qquad\qquad = 13$

(2) $\qquad\qquad\quad x_2 + 3x_3 \quad\;\; + \underline{x}_5 \qquad\quad = 17 \qquad\qquad$ (5.3.1)

(3) $\qquad 2x_1 + \; x_2 + \; x_3 \qquad\qquad + \underline{x}_6 = 13,$

$$x_j \geqslant 0, \qquad j = 1, 2, \ldots, 6.$$

Notice that, for ease in calculation, we have written the objective function in an equation form similar to the constraint equations, and only one basic variable appears in one constraint equation. From the discussions in the previous section, we restrict our attention to basic feasible solutions. The most obvious one is $(0, 0, 0, 13, 17, 13)$, that is, let x_1, x_2, and x_3 be equal to zero and let x_4, x_5, and x_6 be the basic variables. The solution is certainly feasible, and gives $z = 0$.[7]

Next, we seek to improve the result. Since each variable is to be nonnegative, we see that the value of z will be increased if we give a positive value to x_1 or x_2 or x_3, and, in fact, it will be increased fastest if we give a positive value to x_3, its coefficient being the most negative. In other words, we wish to move to a different basic feasible solution in which x_3 will now be a basic variable. What should be the *leaving* basic variable? We look at each of the constraint equations to see what is the largest possible value that can be assigned to x_3 without violating the nonnegativity requirement. As x_3 increases, an original basic variable will decrease in value from positive to zero.

In (1), x_3 may be increased indefinitely without affecting x_1, x_2, or x_4.
In (2), x_3 must be $\leqslant 17/3$. If $x_3 > 17/3$, then x_5 will have to be negative for the equation to hold.
In (3), x_3 must be $\leqslant 13$, for otherwise x_6 in that equation will have to be negative.

Therefore $17/3$ is the largest value that can be assigned to x_3 without driving any of the original basic variables to negative values. When we make the substitution $x_3 = 17/3$, x_5 becomes zero and is the variable to leave the basis. Our new basic feasible solution will be of the form $(0, 0, x_3, x_4, 0, x_6)$. A glance at (5.3.1) tells us that, with the nonbasic variables set equal to zero, we can read off the values of the basic variables $x_4 = 13$, $x_5 = 17$, and $x_6 = 13$. The basic variables have, in fact, been underlined for easy recognition. When we now let x_3 be the basic variable to replace x_5, it is no longer easy to tell at a glance the new values of the basic variables. For instance, in equation (3) there are two basic (nonzero) variables, x_3 and x_6. In order to have the requirements restated in the form

---

[7] To avoid clutter, we will omit vector notation and will not use the superscript T to indicate the transpose of a column vector.

of (5.3.1), we divide equation (2) by 3 to get

(2′) $$\frac{x_2}{3} + x_3 + \frac{x_5}{3} = \frac{17}{3}.$$

Then we eliminate x_3 from equation (3) by the Gauss–Jordan method, that is, multiply each of the two equations by a suitable constant and add the results to eliminate a variable. Here we multiply equation (2′) by -1 and add to equation (3) to get

(3′) $$2x_1 + \frac{2x_2}{3} \qquad - \frac{x_5}{3} + x_6 = \frac{22}{3}.$$

Since

$$x_3 = \frac{17}{3} - \frac{x_2}{3} - \frac{x_5}{3},$$

equation (0) becomes

(0′) $$z - x_1 - 3x_2 - 4\left(\frac{17}{3} - \frac{x_2}{3} - \frac{x_5}{3}\right) = 0.$$

The above work yields the following system which is equivalent to (5.3.1):

(0′) $$z - x_1 - \frac{5}{3}x_2 \qquad\qquad + \frac{4}{3}x_5 \qquad = \frac{68}{3};$$

(1′) $$3x_1 + 2x_2 + 0x_3 + \underline{x_4} + 0x_5 + 0x_6 = 13;$$

(2′) $$0x_1 + \frac{1}{3}x_2 + \underline{x_3} + 0x_4 + \frac{1}{3}x_5 + 0x_6 = \frac{17}{3}; \qquad (5.3.2)$$

(3′) $$2x_1 + \frac{2}{3}x_2 + 0x_3 + 0x_4 - \frac{1}{3}x_5 + \underline{x_6} = \frac{22}{3}.$$

The new basic variables are underlined. The new basic feasible solution is $(0, 0, 17/3, 13, 0, 22/3)$, and at that point $z = 68/3$, an improvement over the first solution.

Is this second basic feasible solution optimal? Looking at (0′), the objective function z may be increased further by assigning positive values to the two nonbasic variables x_1 and x_2 since their coefficients are negative. Increasing x_2 will increase z faster because its coefficient, $-5/3$, is larger in absolute value than the coefficient, -1, of x_1. So, we repeat the procedure. Choose x_2 as the new *entering* basic variable, and examine the constraint equations to decide on the greatest increase in x_2 allowable so that no basic variable will decrease to a negative value.

	Equation	Increase in x_2
(1′)	$x_4 = 13 - 3x_1 - 2x_2,$	$x_2 \leqslant \frac{13}{2}$
(2′)	$x_3 = \frac{17}{3} - \frac{1}{3}x_2 - \frac{1}{3}x_5,$	$x_2 \leqslant 17$
(3′)	$x_6 = \frac{22}{3} - 2x_1 - \frac{2}{3}x_2 + \frac{1}{3}x_5$	$x_2 \leqslant 11$

Therefore the greatest value we can assign to x_2 is $13/2$, at which value $x_4 = 0$ and leaves the basis.

We use the Gauss–Jordan process of elimination again to obtain the equivalent system:

(0″) $\qquad z + \frac{3}{2}x_1 \qquad\qquad\quad + \frac{5}{6}x_4 + \frac{4}{3}x_5 \qquad = \frac{67}{2};$

(1″) $\qquad \frac{3}{2}x_1 + x_2 + 0x_3 + \frac{1}{2}x_4 + 0x_5 + 0x_6 = \frac{13}{2};$ $\qquad\qquad$ (5.3.3)

(2″) $\qquad -\frac{1}{2}x_1 + 0x_2 + x_3 - \frac{1}{6}x_4 + \frac{1}{3}x_5 + 0x_6 = \frac{7}{2};$

(3″) $\qquad x_1 + 0x_2 + 0x_3 - \frac{1}{3}x_4 - \frac{1}{3}x_5 + x_6 = 3.$

The new basic feasible solution is $(0, 13/2, 7/2, 0, 0, 3)$, at which point $z = 67/2$. Is this solution optimal? The answer is yes, because the coefficients of the nonbasic variables in (0″) are all positive. Increasing any of these variables from 0 to a positive value will only decrease the value of the objective function. We conclude that the value of z cannot be increased beyond $67/2$.

The above procedure may be applied in general to any maximization problem with "less than" inequality constraints, and we summarize the steps as follows:

Step 1. Introduce a slack variable for each of the m inequalities and select the m slack variables as the initial basic variables. When doing so, the origin becomes the initial basic feasible solution.

Step 2. To determine the new (entering) basic variable, choose the nonbasic variable with the most negative (largest in absolute value) coefficient in the objective function. This variable, when increased, would increase z fastest.

Step 3. To determine the leaving basic variable, choose the one which decreases and reaches zero first as the entering variable is increased. Since the constraint equations are written in such a way that the constant terms b_i give the values of the initial basic variables, all we need to do is divide the constant term of each equation by the coefficient of the entering variable (if the coefficient > 0) and choose the basic variable in the equation with the smallest ratio as the leaving variable. For instance, from (5.3.2) we get the ratios $13/2$ for (1′), $(17/3)/(1/3) = 17$ for (2′), $(22/3)/(2/3) = 11$ for (3′), and we choose x_4 as the leaving variable because it is the one which becomes zero first (as x_2 becomes $13/2$). If the coefficient of the entering variable is either 0 (as in equation (1) of (5.3.1), where the coefficient of the entering variable x_3 is 0) or negative, an increase of the value of that entering variable will not decrease the value of the current basic variable, so that we only need to consider the positive ratios.[8]

We have completed the change of basis process and move next to the pivoting process.

---

[8]The case where the constant b_i is nonpositive will be considered in a later section.

Step 4. Replace each equation by a new one to reflect the new basic variables, still maintaining the form that each constraint equation contains only one basic variable with its coefficient equal to 1. This may be accomplished by the following procedure. First, call the (horizontal) row containing the leaving basic variable the *pivot row*, the coefficient of the entering basic variable in the pivot row the *pivot element*, and the (vertical) column containing the coefficients of the entering basic variable in all the equations the *pivot column*. For example, in (5.3.2), when we decide that x_2 is to be the entering variable and x_4 the leaving, the pivot row is equation (1'), the pivot element is 2, and the pivot column is the column containing the coefficients of x_2 in the four equations, namely, $-5/3$, 2, $1/3$, and $2/3$. The procedure is:

1. Divide each term of the pivot row by the pivot element to get new row (1''). This makes the coefficient of the newly entered basic variable equal to 1.
2. For any other row, say row k ($k \neq$ the pivot row number), look at the coefficient in the pivot column. Multiply the new row (1'') by the negative of that coefficient. Add to row k. This results in 0 becoming the coefficient of the new basic variable in this row k.

Still using (5.3.2) as illustration, with the pivot element$=2$, (1'), the pivot row, becomes

(1'') $$\tfrac{3}{2}x_1 + x_2 + 0x_3 + \tfrac{1}{2}x_4 + 0x_5 + 0x_6 = \tfrac{13}{2}.$$

To change row (0'), multiply (1'') by $5/3$ to get

$$\tfrac{5}{2}x_1 + \tfrac{5}{3}x_2 + 0x_3 + \tfrac{5}{6}x_4 + 0x_5 + 0x_6 = \tfrac{65}{6}.$$

Add this to row (0'), and get row

(0'') $$z + \tfrac{3}{2}x_1 + 0x_2 + 0x_3 + \tfrac{5}{6}x_4 + 0x_5 + 0x_6 = \tfrac{67}{2}.$$

To change row (2'), multiply (1'') by $-1/3$ to get

$$-\tfrac{1}{2}x_1 - \tfrac{1}{3}x_2 + 0x_3 - \tfrac{1}{6}x_4 + 0x_5 + 0x_6 = -\tfrac{13}{6}.$$

Add this to row (2') and get row

(2'') $$-\tfrac{1}{2}x_1 + 0x_2 + x_3 - \tfrac{1}{6}x_4 + \tfrac{1}{3}x_5 + 0x_6 = \tfrac{7}{2}.$$

Similarly for row (3'').

Step 5. To determine whether this basic feasible solution is optimal, examine the new equation (0), which contains the objective function but no basic variable. If now the coefficients of the variables (nonbasic) are all positive, then z cannot be increased any further and an optimal solution has been obtained. If some of the coefficients are negative, return to Step 2 and repeat the iteration.

To facilitate computation, one usually uses a simplex tableau which records the essential numbers (a_{ij}, coefficients of the variables, and constant terms b_i). There are, of course, various forms that the tableau may take, and one such form is as shown in Table 5.3.1. In it we label the

Table 5.3.1. Simplex tableau

Iteration	Basis	Current coefficient						Current value	Ratio
		x_1	x_2	x_3	x_4	x_5	x_6		
1	z	-1	-3	-4	0	0	0	0	
	x_4	3	2	0	1	0	0	13	$\frac{13}{0}$
	x_5	0	1	$\underline{3}$	0	1	0	17	$\frac{17}{3}$(min)
	x_6	2	1	1	0	0	1	13	$\frac{13}{1}$
2	z	-1	$-\frac{5}{3}$	0	0	$\frac{4}{3}$	0	$\frac{68}{3}$	
	x_4	3	$\underline{2}$	0	1	0	0	13	$\frac{13}{2}$(min)
	x_3	0	$\frac{1}{3}$	1	0	$\frac{1}{3}$	0	$\frac{17}{3}$	$\frac{17}{3}\div\frac{1}{3}$
	x_6	2	$\frac{2}{3}$	0	0	$-\frac{1}{3}$	1	$\frac{22}{3}$	$\frac{22}{3}\div\frac{2}{3}$
	z	$\frac{3}{2}$	0	0	$\frac{5}{6}$	$\frac{4}{3}$	0	$\frac{67}{2}$	Optimal value
	x_2	$\frac{3}{2}$	1	0	$\frac{1}{2}$	0	0	$\frac{13}{2}$	Optimal basic
	x_3	$-\frac{1}{2}$	0	1	$-\frac{1}{6}$	$\frac{1}{3}$	0	$\frac{7}{2}$	feasible solution
	x_6	1	0	0	$-\frac{1}{3}$	$-\frac{1}{3}$	1	3	

objective function z a basic variable for convenience. The pivot element for each iteration is underlined.

Looking back, we see that the first basic feasible solution corresponds to the origin, or point O, in Figure 5.2.5, the second to the point D, and the third and last to the point C. Another noticeable feature of the simplex method is that it replaces only one basic variable at a time, thus moving from one extreme point to a "neighboring" point which yields a higher value for the objective function.

The above procedure was written with a maximization problem in mind. If we wish to minimize a function $f=\sum_{j=1}^{n}c_j x_j + d$, we can multiply each term in the objective function by -1 and try, instead, to maximize $-f$, since minimizing any function $f(x_1,x_2,\ldots,x_n)$ is equivalent to maximizing $-f(x_1,x_2,\ldots,x_n)$, subject to the same constraints. Or, we can attack the minimization problem directly by using the simplex method, this time keeping in mind that the entering basic variable should be the one which decreases the value of the objective function, z, fastest. For example, to minimize z in

$$z - 3x_1 + 4x_2 - 5x_3 + 6x_4 = 0,$$

choose x_4 as the entering basic variable. When testing for optimality, we also see whether the value of z may be decreased by increasing any nonbasic variable. If the constraint inequalities are in the same direction ($\leqslant b_i$) as illustrated, no further revision in the method is needed, because Steps 3 and 4 are aimed solely at keeping the new basic solution feasible and easy to read.

In actual practice, however, we will find that the inequalities are usually in the opposite direction. When this is the case, there is no longer an obvious initial basic feasible solution, in contrast to the maximization problem in which the origin often serves the purpose. An example may bring out the idea. Minimize

$$z = x_1 + 3x_2$$

subject to

$$4x_1 + x_2 \geqslant 8,$$

$$x_1 + x_2 \geqslant 5,$$

$$x_1 - 3x_2 \geqslant 1,$$

$$x_1 \geqslant 0, \qquad x_2 \geqslant 0.$$

As Figure 5.3.1 shows, the origin is not in the feasible region. If we introduce surplus variables to the constraints to get

$$4x_1 + x_2 - x_3 \qquad\qquad = 8,$$

$$x_1 + x_2 \qquad - x_4 \qquad = 5,$$

$$x_1 - 3x_2 \qquad\qquad - x_5 = 1,$$

$$x_i \geqslant 0, \qquad i = 1, 2, \ldots, 5,$$

and set x_1 and x_2 equal to 0 (so that they are nonbasic), we would get $(0, 0, -8, -5, -1)$ as a basic solution which is not feasible.

Notice from Figure 5.3.1 that the first constraint is superfluous.

To get around the difficulty, one modifies the simplex technique slightly and introduces artificial variables. We postpone the discussion to the next section. There is another, different approach to this type of minimization problem which involves the concept of duality. We shall look at duality in the next chapter.

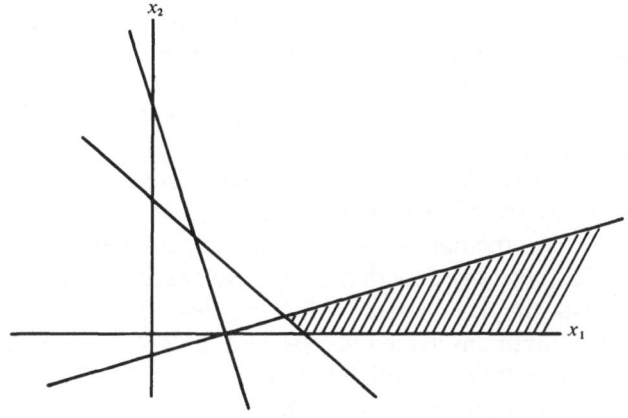

Figure 5.3.1.

5.4 Complications and adjustments

The previous section may lead the reader to the comfortable but false impression that an optimal solution may be reached very easily by simply following the five steps listed. There are, however, other legitimate forms of the linear programming problem which differ slightly from the example presented in Sections 5.2 and 5.3, and we already saw that if the inequalities were in the $\geqslant$ direction, we couldn't even find an obvious initial feasible solution to get the simplex algorithm started. Fortunately, some minor adjustments of the procedure will take care of this and some other complications. We shall take a brief look at them.

5.4.1 Constraint is an equality

Suppose we take the example of Section 5.2 and Section 5.3 and make just one change, by letting the third constraint be $2x_1 + x_2 + x_3 = 13$, instead of $\leqslant 13$. We are then no longer justified in introducing a slack variable x_6 for this constraint, and this situation leaves us with one less initial basic variable. If we are to allow, say, x_3 to be the candidate, we see that the second constraint will become

$$-6x_1 - 2x_2 + 0x_3 + 0x_4 + x_5 = -22$$

and x_5 can no longer be a basic feasible variable since it is equal to a negative number. We will run into similar difficulties if we allow x_1 or x_2 to be the initial basic variable.

As we pointed out in Section 5.1, the equation can be replaced by the two inequalities

$$2x_1 + x_2 + x_3 \leqslant 13 \quad \text{and} \quad 2x_1 + x_2 + x_3 \geqslant 13.$$

The second inequality, with the sense directed $\geqslant$, will also lead to a negative basic variable. This case will be discussed later in this section.

Another way to tackle the difficulty is to introduce an artificial nonnegative variable x_6 as if it were a slack variable for an inequality. This changes, of course, the original constraint, although it does enable us to begin with an initial basic feasible solution $x_1 = x_2 = x_3 = 0$, $x_4 = 13$, $x_5 = 17$, $x_6 = 13$. The revised problem would be identical with the original problem if $x_6 = 0$. Therefore, after we have obtained an initial basic feasible solution to the revised problem, we can seek to return to the original problem by driving x_6 to zero. There are two popular ways to achieve this purpose, the Two Phase method and the Big M method.

In the Two Phase method, we first let the objective function be $w = x_6$ and minimize this objective function, subject to the same set of constraints and by the usual simplex method. Since x_6, like the other variables, is nonnegative, the minimum value must be zero, unless the original problem admits no feasible solution at all. This optimal solution for Phase I with $w = 0$ must be a basic feasible solution for the original problem. We then proceed to Phase II by solving the original problem again by the simplex method and using the optimal solution for Phase I as the initial basic

Table 5.4.1

Basis	Current coefficient						Current value	Ratio
	x_1	x_2	x_3	x_4	x_5	x_6		
w	2	1	1	0	0	0	13	
x_4	3	2	0	1	0	0	13	$\frac{13}{3}$
x_5	0	1	3	0	1	0	17	
x_6	2	1	1	0	0	1	13	$\frac{13}{2}$
w	0	$-\frac{1}{3}$	1	$-\frac{2}{3}$	0	0	$\frac{13}{3}$	
x_1	1	$\frac{2}{3}$	0	$\frac{1}{3}$	0	0	$\frac{13}{3}$	
x_5	0	1	3	0	1	0	17	$\frac{17}{3}$
x_6	0	$-\frac{1}{3}$	1	$-\frac{2}{3}$	0	1	$\frac{13}{3}$	$\frac{13}{3}$
w	0	0	0	0	0	-1	0	
x_1	1	$\frac{2}{3}$	0	$\frac{1}{3}$	0	0	$\frac{13}{3}$	Optimal
x_5	0	2	0	2	1	-3	4	solution
x_3	0	$-\frac{1}{3}$	1	$-\frac{2}{3}$	0	1	$\frac{13}{3}$	for Phase I

feasible solution. Let us illustrate the procedure by solving the problem.

Phase I: Minimize

$$w + 2x_1 + x_2 + x_3 + 0x_4 + 0x_5 + 0x_6 - 13 = 0$$

subject to

$$3x_1 + 2x_2 + 0 + x_4 + 0 + 0 = 13,$$

$$0 + x_2 + 3x_3 + 0 + x_5 + 0 = 17,$$

$$2x_1 + x_2 + x_3 + 0 + 0 + x_6 = 13,$$

$$x_i \geqslant 0, \qquad i = 1, \ldots, 6.$$

We summarize the computations in Table 5.4.1.

Phase II: We next take the solution for Phase I, namely, $x_1 = 13/3$, $x_3 = 13/3$[9], $x_5 = 4$, and use these three variables as the initial basic variables for the original problem. Maximize

$$z - \left(\frac{13}{3} - \frac{2}{3}x_2 - \frac{1}{3}x_4\right) - 3x_2 - 4\left(\frac{13}{3} + \frac{1}{3}x_2 + \frac{2}{3}x_4 - x_6\right) = 0,$$

that is,

$$z + 0x_1 - \frac{11}{3}x_2 + 0x_3 - \frac{7}{3}x_4 + 0x_5 + 4x_6 - \frac{65}{3} = 0.$$

The computations are summarized in Table 5.4.2.

[9]Actually, $x_1 = (13/3) - (2/3)x_2 - 0x_3 - (1/3)x_4 - 0x_5 - 0x_6$, and $x_3 = (13/3) + 0x_1 + (1/3)x_2 + (2/3)x_4 + 0x_5 - x_6$, but $x_2 = x_4 = x_6 = 0$.

Table 5.4.2

Basis	Current coefficient						Current value	Ratio
	x_1	x_2	x_3	x_4	x_5	x_6		
z	0	$-\frac{11}{3}$	0	$-\frac{7}{3}$	0	4	$\frac{65}{3}$	
x_1	1	$\frac{2}{3}$	0	$\frac{1}{3}$	0	0	$\frac{13}{3}$	$\frac{13}{2}$
x_5	0	2	0	2	1	-3	4	$\frac{4}{2}$
x_3	0	$-\frac{1}{3}$	1	$-\frac{2}{3}$	0	1	$\frac{13}{3}$	
z	0	0	0	$\frac{4}{3}$	$\frac{11}{6}$	$-\frac{3}{2}$	29	
x_1	1	0	0	$-\frac{1}{3}$	$-\frac{1}{3}$	1	3	Optimal
x_2	0	1	0	1	$\frac{1}{2}$	$-\frac{3}{2}$	2	solution
x_3	0	0	1	$-\frac{1}{3}$	$\frac{1}{6}$	$\frac{1}{2}$	5	for problem

Thus the optimal solution is $x_1=3$, $x_2=2$, $x_3=5$, with $z=29$. Notice that we terminate the procedure although the coefficient of x_6 is $-3/2$ in the first row of the final iteration. Since x_6 is an artificial variable and does not really belong to the original problem, we cannot increase the value of z by increasing the value of x_6 from 0 to a positive number.

Another popular way to solve the problem is the so-called Big M method. Again we introduce an artificial variable x_6 to the constraint. But instead of using the simplex method to drive x_6 to zero, this time we assign a very large penalty coefficient to the artificial variable. If we change the original objective function from

$$z - x_1 - 3x_2 - 4x_3 = 0$$

to

$$z - x_1 - 3x_2 - 4x_3 + Mx_6 = 0,$$

where M is a very large positive number, then the maximum value of z will occur when $x_6=0$ and the optimal solution to the revised problem with the artificial variable will also be the feasible, optimal solution to the original problem. We illustrate the Big M procedure below.

Since the artificial variable x_6 is going to be used as an initial basic variable, it should be removed from the objective function by assigning 0 as its coefficient, in accordance with the simplex method. This can be accomplished by subtracting M times the third constraint from the objective function equation to get: Maximize

$$z - (2M+1)x_1 - (M+3)x_2 - (M+4)x_3 + 13M = 0$$

subject to

$$3x_1 + 2x_2 + 0 + \underline{x_4} + 0 + 0 = 13,$$

$$0 + x_2 + 3x_3 + 0 + \underline{x_5} + 0 = 17,$$

$$2x_1 + x_2 + x_3 + 0 + 0 + \underline{x_6} = 13.$$

Table 5.4.3

Basis	Current coefficient						Current value	Ratio
	x_1	x_2	x_3	x_4	x_5	x_6		
z	$-(2M+1)$	$-(M+3)$	$-(M+4)$	0	0	0	$-13M$	
x_4	3	2	0	1	0	0	13	$\frac{13}{3}$
x_5	0	1	3	0	1	0	17	
x_6	2	1	1	0	0	1	13	$\frac{13}{2}$
z	0	$\frac{M-7}{3}$	$-(M+4)$	$\frac{2M+1}{3}$	0	0	$\frac{-13M+17}{3}$	
x_1	1	$\frac{2}{3}$	0	$\frac{1}{3}$	0	0	$\frac{13}{3}$	
x_5	0	1	3	0	1	0	17	$\frac{17}{3}$
x_6	0	$-\frac{1}{3}$	1	$-\frac{2}{3}$	0	1	$\frac{13}{3}$	$\frac{13}{3}$
z	0	$\frac{11}{3}$	0	$-\frac{7}{3}$	0	$M+4$	$\frac{65}{3}$	
x_1	1	$\frac{2}{3}$	0	$\frac{1}{3}$	0	0	$\frac{13}{3}$	$\frac{13}{2}$
x_5	0	2	0	2	1	-3	4	2
x_3	0	$-\frac{1}{3}$	1	$-\frac{2}{3}$	0	1	$\frac{13}{3}$	
z	0	0	0	$\frac{4}{3}$	$\frac{11}{6}$	$M-\frac{3}{2}$	29	
x_1	1	0	0	$-\frac{1}{3}$	$-\frac{1}{3}$	1	3	Optimal
x_2	0	1	0	1	$\frac{1}{2}$	$-\frac{3}{2}$	2	solution
x_3	0	0	1	$-\frac{1}{3}$	$\frac{1}{6}$	$\frac{1}{2}$	5	

The computations are summarized in Table 5.4.3. The optimal solution, of course, is still $x_1=3$, $x_2=2$, $x_3=5$, with $z=29$.

Since the Big M method is essentially equivalent to the Two Phase method in that they both seek to drive the artificial variable to zero, it should be no surprise that they have the same sequence of basic feasible solutions in their iterations.

Notice that the optimal solution to this problem corresponds to the point G of Figure 5.2.5.

If more than one constraint is stated as an equality so that several artificial variables (one for each equality) are involved, we can add all the artificial variables to save some work and, in Phase I of the Two Phase method, minimize the sum of the artificial variables. In the Big M method, we multiply each constraint equation which contains an artificial variable by a very large M and subtract each revised constraint equation from the revised objective function equation. If the original problem is a minimization one, a suitable adjustment is made by adding, instead of subtracting, M times the constraint equation with an artificial variable to the revised objective function equation.

5.4.2 Nonpositive b_i

A constraint may come with a negative constant term b_i, or the b_i may become negative as a result of changing an inequality from $\geqslant$ to $\leqslant$. For instance, if an inequality $3x_1 + 2x_2 \geqslant 13$ is changed to $-3x_1 - 2x_2 \leqslant -13$ as suggested in Section 5.1, the addition of a slack variable will not help matters immediately. Consider our problem of Section 5.3 again, in which the first constraint is changed to $\geqslant$ while keeping the others unchanged:

$$
\begin{aligned}
z - x_1 - 3x_2 - 4x_3 \qquad\qquad\qquad &= 0; \\
-3x_1 - 2x_2 \qquad + x_4 \qquad\qquad &= -13; \\
x_2 + 3x_3 \qquad + x_5 \qquad &= 17; \\
2x_1 + x_2 + x_3 \qquad\qquad + x_6 &= 13.
\end{aligned}
$$

The initial basic solution of $(0, 0, 0, -13, 17, 13)$ is not feasible because $x_4 = -13$. What is commonly done is to subtract a nonnegative artificial variable x_7 from this constraint so that the equation becomes

$$-3x_1 - 2x_2 + x_4 - x_7 = -13,$$

that is,

$$3x_1 + 2x_2 - x_4 + x_7 = 13.$$

This artificial variable will then be used as an initial basic variable together with x_5 and x_6. Of course, the original problem has been revised. By adding x_4 and subtracting x_7, each of which can be any nonnegative number, we have in effect nullified this constraint. In so doing, we may have enlarged the original set of feasible solutions. To take care of this complication, we drive the artificial variable to zero, and the methods discussed previously may again be applied. To illustrate the Big M method in this case, the objective function, first appearing as

$$z - x_1 - 3x_2 - 4x_3 + Mx_7 = 0,$$

becomes

$$z - (3M + 1)x_1 - (2M + 3)x_2 - 4x_3 + Mx_4 = -13M$$

after we subtract M times the first constraint from it. The remaining computations proceed as usual.

We have discussed the situation with $b_i < 0$. Suppose $b_i = 0$ for some i. This is what is called the *degenerate case* and we look at it in the next paragraph.

5.4.3 Degeneracy

Degeneracy is indicated at a given stage of the simplex procedure when a basic variable takes on a zero value. This may come about in two ways: Some b_i may be 0 in an original constraint; or, at some iteration in the simplex algorithm two or more variables reach zero simultaneously as the entering basic variable is increased. As illustration, let us modify our problem of Section 5.3 this time by changing the second constraint from $x_2 + 3x_3 \leqslant 17$ to $x_2 + 3x_3 \leqslant 39$. The initial set of equations is:

$$z - x_1 - 3x_2 - 4x_3 \qquad\qquad = 0;$$
$$3x_1 + 2x_2 \qquad + x_4 \qquad\qquad = 13;$$
$$x_2 + 3x_3 \qquad + x_5 \qquad = 39;$$
$$2x_1 + x_2 + x_3 \qquad\qquad + x_6 = 13.$$

As before, x_3 is chosen as the entering basic variable, but both x_5 and x_6 reach 0 as x_3 is increased to 13. Suppose that we arbitrarily choose x_5 as the leaving variable. In the next iteration we find the following set of equations:

$$z - x_1 - \tfrac{5}{3}x_2 \qquad\qquad + \tfrac{4}{3}x_5 \qquad = 52;$$
$$3x_1 + 2x_2 \qquad + x_4 \qquad\qquad = 13;$$
$$\tfrac{1}{3}x_2 + x_3 \qquad + \tfrac{1}{3}x_5 \qquad = 13;$$
$$2x_1 + \tfrac{2}{3}x_2 \qquad\qquad - \tfrac{1}{3}x_5 + x_6 = 0.$$

The corresponding basic feasible solution of $(0, 0, 13, 13, 0, 0)$ is degenerate because a basic variable, x_6, is zero and it appears to be not optimal either because both x_1 and x_2 may be increased. According to the simplex algorithm, we increase x_2. Since it can be increased without driving x_6 negative, we get the next set of equations:

$$z + 4x_1 \qquad\qquad + \tfrac{1}{2}x_5 + \tfrac{5}{2}x_6 = 52;$$
$$-3x_1 \qquad\qquad + x_4 + x_5 - 3x_6 = 13;$$
$$-x_1 \qquad + x_3 \qquad + \tfrac{1}{2}x_5 - \tfrac{1}{2}x_6 = 13;$$
$$3x_1 + x_2 \qquad\qquad - \tfrac{1}{2}x_5 + \tfrac{3}{2}x_6 = 0.$$

The solution $(0, 0, 13, 13, 0, 0)$ with $z = 52$ is the same, but this time we find that it is optimal.

One is not always that lucky in reaching an optimal solution so quickly when degeneracy occurs. A famous example by Beale (1955) shows that *cycling* is a possibility. That is, it is possible that when several of the ratios b_i / a_{ipc} (where a_{ipc} stands for the coefficient in the ith row and pivot column) *tie* for being the smallest and we arbitrarily choose one of the tied variables as the entering one, we may find ourselves returning some iterations later to a basic solution which we have encountered before. This basic sequence may be repeated again and again with no improvement of the objective function. Fortunately, cycling is extremely rare and there are several ways to prevent cycling to reach the optimum. The interested reader is referred to: Charnes (1952): Dantzig (1952): Garvin (1960), Chapter 14; and Spivey and Thrall (1970), pp. 90–99.

If a basic variable is equal to zero, this means that **b**, the column vector of constants of (5.2.8), can be expressed as a linear combination of fewer than m linearly independent vectors. To see this, recall Cramer's rule for solving a system of m equations in m unknowns in which the solution of a variable is

$$x_j = \frac{\begin{vmatrix} a_{11}\ldots a_{1j-1} b_1\, a_{1j+1}\ldots a_{1m} \\ \vdots \\ a_{m1}\ldots a_{mj-1} b_m\, a_{mj+1}\ldots a_{mm} \end{vmatrix}}{\det A} \tag{5.4.1}$$

where $\det A$ means the determinant of the matrix of coefficients a_{ij} of the m unknown variables. We also recall from algebra that if a determinant is equal to 0, some column may be expressed as a linear combination of the remaining columns. Looking at (5.4.1), several possibilities occur:

(1) Det A, the denominator determinant, is not 0. The coefficients of the unknowns are linearly independent and the system has a unique solution. If the numerator determinant is 0, the column of constants is a linear combination of the remaining $m-1$ columns.

(2) Det $A = 0$ and the numerator determinant is not 0. Although there is linear dependence among the coefficients a_{ij}, the constant column is independent of any $m-1$ columns of the a_{ij}'s. The system is said to be inconsistent and there is no solution.

(3) Det $A = 0$ and the numerator determinant is also 0. The coefficients are linearly dependent and the constant column is linearly dependent on the coefficients of $m-1$ unknowns. The system may have an infinite number of solutions, with some variable(s) expressed as a linear combination of the others.

Linear dependency implies that at least one of the m equations is redundant (hence it can be expressed as a combination of the others). Returning to degeneracy in linear programming, this means that some constraint is redundant; it is a linear combination of some other constraints. When we introduce a slack or artificial variable to each constraint equation as we do in the simplex method, we also set the nonbasic variables equal to zero and assign a coefficient of 1 to each basic variable. This prevents the denominator from becoming zero but will not prevent the numerator determinant, hence a zero value for some basic variable.

Graphically, degeneracy is indicated by the intersection of more than n hyperplanes at a corner or extreme point. In two-space, only two hyperplanes (lines) are needed to meet at a point; in three-space, three hyperplanes (planes) are needed; and so on. Let us illustrate the situation with two examples in two-space. Suppose we wish to maximize

$$z = x_1 + 3x_2$$

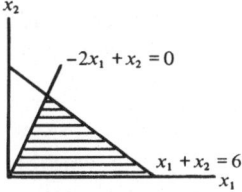

Figure 5.4.1.

subject to

$$x_1 + x_2 \leqslant 6,$$

$$-2x_1 + x_2 \leqslant 0,$$

$$x_1 \geqslant 0, \quad x_2 \geqslant 0.$$

As usual, we introduce slack variables x_3 and x_4 to change the inequalities to equations. But we find that in the initial basic feasible solution $(0,0,6,0)$, the basic variable $x_4 = 0$. As seen in Figure 5.4.1, three lines, instead of two, meet at the origin. The requirement $x_1 \geqslant 0$ is redundant, because this follows from the two requirements $x_2 \geqslant 0$ and $-2x_1 + x_2 \leqslant 0$.

As a second example, again maximize

$$z = x_1 + 3x_2$$

but now subject to

$$x_1 + x_2 \leqslant 6, \qquad 2x_1 + x_2 \leqslant 8, \qquad 3x_1 + x_2 \leqslant 10, \qquad x_1 \geqslant 0, \qquad x_2 \geqslant 0.$$

Figure 5.4.2 shows that the second constraint is redundant. This example also illustrates another interesting feature. The point B is a degenerate extreme point, and the maximum value for the objective function z can be easily shown to occur at the vertex C. By the simplex method, we would enter x_2 to replace x_3 as the basic variable and get to point C immediately from the point O. If we do this, we would not even be aware of the degeneracy. On the other hand, if for some reason we should choose x_1, instead of x_2, as the entering basic variable, then we would be traveling from point O to point A, then to point B. At point B we will discover the

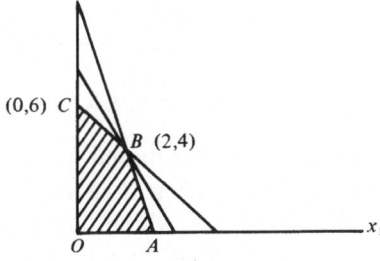

Figure 5.4.2.

degeneracy, but will not be trapped there and will still be able to move on to point C. The reader is urged to try the problem as an exercise.

5.4.4 Tie for entering basic variable

It may happen that several nonbasic variables in the objective function have the same largest coefficient which will increase z fastest. This poses no complication at all and one may select any of these tied variables as the entering basic variable.

5.4.5 Variables unrestricted in sign

Although the decision variables x_j are required to be nonnegative in most cases, this is not always or necessarily so. As we said in Section 5.1, we can write x_j as the difference of two nonnegative variables as $x_j = x_j' - x_j''$. Since the simplex method examines basic feasible solutions, or extreme points, only, these two variables will not both assume positive values at the same time. The same technique suggested in Section 5.1 for dealing with several unrestricted or lower bounded variables may also be used without creating any difficulties in the simplex method.

5.4.6 Multiple solutions

Suppose we add a variable x_7 to our example so that (5.3.1) now reads:
Maximize

$$(0) \qquad z - x_1 - 3x_2 - 4x_3 \qquad\qquad + \quad 2x_7 = 0$$

subject to

$$(1) \qquad 3x_1 + 2x_2 \qquad + x_4 \qquad\qquad - \tfrac{36}{5}x_7 = 13,$$

$$(2) \qquad\qquad x_2 + 3x_3 \qquad + x_5 \qquad + \quad 3x_7 = 17, \qquad (5.4.2)$$

$$(3) \qquad 2x_1 + x_2 + x_3 \qquad\qquad + x_6 \qquad\qquad = 13.$$

The reader can verify that the simplex method will lead to the final set of equations

$$(0) \qquad z + \tfrac{3}{2}x_1 \qquad\qquad + \tfrac{5}{6}x_4 + \tfrac{4}{3}x_5 \qquad\qquad = \tfrac{67}{2},$$

$$(1) \qquad \tfrac{3}{2}x_1 + x_2 \qquad + \tfrac{1}{2}x_4 \qquad\qquad - \tfrac{18}{5}x_7 = \tfrac{13}{2}, \qquad (5.4.3)$$

$$(2) \qquad -\tfrac{1}{2}x_1 \qquad + x_3 - \tfrac{1}{6}x_4 + \tfrac{1}{3}x_5 \qquad + \tfrac{11}{5}x_7 = \tfrac{7}{2},$$

$$(3) \qquad x_1 \qquad\qquad - \tfrac{1}{3}x_4 - \tfrac{1}{3}x_5 + x_6 + \tfrac{7}{5}x_7 = 3.$$

This value of the objective function is optimal and the optimal solution is $\mathbf{x}^{o_1} = (0, \tfrac{13}{2}, \tfrac{7}{2}, 0, 0, 3, 0)^{\mathrm{T}}$. Now the coefficient of each nonbasic variable in equation (0) indicates the extent to which the value of the objective function z may be increased or decreased when that variable is introduced as a basic variable. Since the coefficient of the nonbasic variable x_7 is zero in equation (0), entering that into the basis will neither increase nor decrease the value of z. Let us enter x_7 as a basic variable, and get, according to the simplex method, these equations:

(0) $z + \frac{3}{2}x_1 \qquad\qquad + \frac{5}{6}x_4 + \frac{4}{3}x_5 \qquad\qquad = \frac{67}{2}$;

(1) $\frac{15}{22}x_1 + x_2 + \frac{18}{11}x_3 + \frac{5}{22}x_4 + \frac{6}{11}x_5 \qquad\qquad = \frac{269}{22}$;

(2) $-\frac{5}{22}x_1 \qquad + \frac{5}{11}x_3 - \frac{5}{66}x_4 + \frac{5}{33}x_5 \qquad + x_7 = \frac{35}{22}$;

(3) $\frac{29}{22}x_1 \qquad - \frac{7}{11}x_3 - \frac{5}{22}x_4 - \frac{6}{11}x_5 + x_6 \qquad = \frac{17}{22}$.

As expected, the value of the objective function remains at $67/2$, but the new set of optimal solutions is $\mathbf{x}^{o_2} = (0, \frac{269}{22}, 0, 0, 0, 17/22, 35/22)^T$. Thus we have two optimal feasible solutions. Since the set of feasible solutions is a convex set, any weighted average $k\mathbf{x}^{o_1} + (1-k)\mathbf{x}^{o_2}$, $0 \leqslant k \leqslant 1$, is also a feasible optimal solution. That is to say, we have an infinite number of optimal solutions all yielding the same value for the objective function. The simplex method stops as soon as one optimal solution is reached. But by looking at the coefficients of the nonbasic variables in the final objective function equation (0), we can tell that there are alternate optimal solutions if one of these coefficients is zero.

5.4.7 Unbounded optimal solution

It is possible that the nature of a problem is such that some nonbasic variable(s) may be increased indefinitely without violating feasibility. The simplex method will show that when such a variable is chosen as an entering basic variable, it will not drive any present basic variable negative; hence there is no leaving basic variable. We may assign any arbitrarily large value to that nonbasic variable, and the objective function in turn becomes arbitrarily large. Figure 5.2.2 of Section 5.2 shows a problem in which the objective function does not have a finite maximum.

5.4.8 No feasible solution

It is also possible that no feasible solution exists. It may happen that we cannot find an obvious initial basic feasible solution, or, if we have to introduce an artificial variable, that the artificial variable refuses to be driven to zero as it is supposed to. The requirements may be incompatible or may not admit a feasible solution. When using the simplex method one should make sure at each step that the restraints are all satisfied.

5.5 Solving a minimization problem

Let us try to minimize

$$z_y = 13y_1 + 17y_2 + 13y_3$$

subject to

$$3y_1 \quad\quad + 2y_3 \geqslant 1,$$
$$2y_1 + y_2 + y_3 \geqslant 3,$$
$$3y_2 + y_3 \geqslant 4,$$
$$y_i \geqslant 0, \quad i = 1, 2, 3.$$

(5.5.1)

When we introduce surplus variables, the system of constraints becomes

$$3y_1 \quad\quad + 2y_3 - y_4 \quad\quad = 1,$$
$$2y_1 + y_2 + y_3 \quad - y_5 \quad = 3,$$
$$3y_2 + y_3 \quad\quad - y_6 = 4,$$
$$y_i \geqslant 0, \quad i = 1, 2, \ldots, 6.$$

(5.5.2)

There is no obvious basic feasible solution because we do not want y_4, y_5, or y_6 to take on negative values. As suggested in the previous section, we introduce artificial variables to the system and use the Two Phase or Big M method. If we use the Big M method, the objective function is first written as

(0) $\quad\quad z_y - 13y_1 - 17y_2 - 13y_3 - My_7 - My_8 - My_9 = 0.$

Notice that since this is a minimization problem the penalty coefficients are assigned arbitrarily large values with negative sign in equation (0). Next we try to assign a coefficient of 0 to the initial basic variables y_7, y_8, and y_9 in accordance with the simplex method. The initial set of equations is:

(0) $z_y = (13 - 5M)y_1 - (17 - 4M)y_2 - (13 - 4M)y_3 - My_4 - My_5 - My_6 = 8M;$

(1) $\quad\quad 3y_1 \quad\quad\quad\quad +2y_3 \ -y_4 \quad\quad + \underline{y_7} = 1;$

(2) $\quad\quad 2y_1 \quad\quad +y_2 \quad\quad + y_3 \quad\quad -y_5 + \underline{y_8} = 3;$

(3) $\quad\quad\quad\quad\quad\quad 3y_2 \quad\quad + y_3 \quad\quad -y_6 + \underline{y_9} = 4.$

The entering basic variable should be the one with the largest positive coefficient in equation (0), namely, the one whose value, when increased, will decrease the value of the objective function fastest. In our problem it is y_1. Aside from the above modifications of the maximization problem, the simplex method proceeds as usual and y_7 should be the leaving basic variable. The final set of equations is:

(0) $\quad z_y - 3y_3 - \frac{13}{2}y_5 - \frac{7}{2}y_6 - My_7 + (-M + \frac{13}{2})y_8 + (-M + \frac{7}{2})y_9 = \frac{67}{2};$

(1) $\quad \underline{y_1} + \frac{1}{3}y_3 - \frac{1}{2}y_5 + \frac{1}{6}y_6 \quad\quad + \frac{1}{2}y_8 \quad\quad - \frac{1}{6}y_9 = \frac{5}{6};$

(2) $\quad -y_3 + \underline{y_4} - \frac{3}{2}y_5 + \frac{1}{2}y_6 \quad -y_7 + \frac{3}{2}y_8 \quad - \frac{1}{2}y_9 = \frac{3}{2};$

(3) $\quad\quad + \underline{y_2} + \frac{1}{3}y_3 - \frac{1}{3}y_6 \quad\quad\quad\quad + \frac{1}{3}y_9 = \frac{4}{3}.$

The coefficients of the nonbasic variables in equation (0) are all negative, so the value of the objective function cannot be decreased any further. The optimal solution is $(5/6, 4/3, 0, 3/2, 0, 0)$, with $z_y = 67/2$.

5.6 Assumptions of linear programming

Aside from the usual, although not necessary, assumption of nonnegativity of the activity variables in the linear programming model, there are certain other assumptions which limit the applications of the technique.

Perhaps the two foremost assumptions of the model are proportionality and additivity. By virtue of linearity of the objective function as well as the constraint relations, it is assumed, then, that: (1) The contribution to the objective function of each activity taken individually is directly proportional to the level of that activity; (2) The use of a factor by each activity taken individually is again directly proportional to the level of that activity. For example, let the objective function be total profit, x_j the amount of commodity j produced, c_j the profit from the sale of each unit of product j, b_i the total amount available of factor i, and a_{ij} the amount of factor i used in the production of product j. Suppose $c_j = \$30$, that is, the profit from one unit of product j is \$30. If we double the output of product j, then the profit from product j also doubles to \$60. Similarly, suppose $a_{ij} = 3$, that is, 3 units of factor i is used in the production of one unit of product j. If $x_j = 10$, 30 units of factor i will be used to produce output j. If x_j is doubled to 20, 60 units of factor i will be used to produce output j. In economics, the constants of proportionality in the constraint functions indicate constant returns to scale. Obviously there are many programming problems which do not justify the assumption of proportionality. Nonlinear programming concerns itself with such situations, but nonlinear programming is applicable only in certain special cases, as will be seen in a later chapter.

Linearity also requires the additivity assumption, which states that total profit is equal to the sum of the profits from the individual activities, and, for the constraint functions, that total resource used is equal to the sum of the resources used for the individual activities. Such an assumption does not allow for interactions between the activities, whereas in actual situations interactions frequently do exist.

Another assumption that limits the use of the model in certain real-life problems is that the activity variables may take on fractional values such as $7/2$. Suppose x_j stands for the number of airplanes manufactured. It would be meaningless to talk about the profit from $7/2$ airplanes. Integer linear programming is designed to meet such objections, but the use of integer programming is also rather limited.

Finally, the constants in the model, namely the c_j's, b_i's, and a_{ij}'s, which we call *parameters*, are assumed to be known with certainty. This again may not be the case in real situations. Even if it is, new information may necessitate changes in the parameters. The next chapter, on sensitivity analysis, will have a brief look at some simple changes. Parametric pro-

gramming is an extension of sensitivity analysis. Sometimes the parameters
are not even constants but are random variables, in which case a probabil-
istic approach called *stochastic programming* or *programming under uncer-
tainty* may be adopted.

Exercises

5.1 Do the problem of Figure 5.4.2 by choosing x_1 to replace x_3 as the basic
variable.

5.2 Verify the equations in (5.4.3).

5.3 Fill in the steps for the minimization problem of Section 5.5.

5.4 The vector equation $v_1 x_1 + \ldots + v_n x_n + e_1 s_1 + \ldots + e_m s_m = b$, where $v_j = (a_{11}, a_{21}, \ldots, a_{m1})^T$, e_i is the unit vector with the ith coordinate $=1$ and 0
elsewhere, s_i is a slack variable, and $b = (b_1, \ldots, b_m)^T$ (for $i = 1, \ldots, m$ and
$j = 1, \ldots, n$), is often associated with activity analysis. Suppose that a firm uses
two resources, capital K and labor L to produce three goods X_1, X_2, and X_3,
and the resources are limited to K_0 and L_0. The vector equation is

$$\begin{pmatrix} a_{11} \\ a_{21} \end{pmatrix} x_1 + \begin{pmatrix} a_{12} \\ a_{22} \end{pmatrix} x_2 + \begin{pmatrix} a_{13} \\ a_{23} \end{pmatrix} x_3 + \begin{pmatrix} 1 \\ 0 \end{pmatrix} s_1 + \begin{pmatrix} 0 \\ 1 \end{pmatrix} s_2 = \begin{pmatrix} K_0 \\ L_0 \end{pmatrix},$$

and we can regard each term as an activity. For instance, the first activity
consists of using a_{11} units of K and a_{21} units of L to produce x_1 units of X_1.
The problem is to determine the optimal levels of each activity. (We may treat
each slack variable as being associated with the activity of leaving some
resource unused.)
    What assumptions regarding a production function are implicit in this
activity analysis framework?

5.5 In classical economic analysis, the graph of the production function of a firm
which uses K and L to produce one output Q is a smooth curve called an
isoquant. What would be a comparable graph for this production function in
the activity analysis framework?

5.6 Suppose the firm employs three processes simultaneously to produce a single
good Q. Let x_j ($j = 1, 2, 3$) be the output process, so that $\sum_{j=1}^{3} x_j = q$, where q is
the output of Q. Each process requires the use of certain amounts of resources
K and L, which are fixed at K_0 and L_0. Let $p = $ market price of Q, $c_j = $ unit cost
of the output of the jth process, $\pi = $ profit. Each process operates under
constant returns to scale independent of the other processes.
    (a) Formulate the problem as one of maximizing profit.
    (b) What is the graph of the isoquant $q\, (=\Sigma x_j) = 1$?

5.7 For the firm described above, the following table gives the requirements of
three processes to produce one unit of Q. Assume that 6 units each of K and L
are available. Find the most efficient use of the processes to maximize output.

Process	Requirements for K	L
A	2	4
B	3.5	3.5
C	4	2

5.8 A steel company is required to reduce its annual emission of the following two pollutants by at least these amounts: sulfur oxides 3 and hydrocarbons 4 (both in million pounds). The firm has two main abatement methods, but both have technological limits on how much of each pollutant (in millions of pounds per year) they can eliminate, as follows.

Pollutant	Method	
	I	II
sulfur oxides	1	1
hydrocarbons	1	2

The cost of using method I is $10 and of method II is $12 (both in millions). Assume that the cost of lesser uses of a method is proportional to its fractional capacity. Determine the optimal combination of the two methods, perhaps with fractional capacities, to meet the antipollution requirements at minimal cost.

6 Linear Programming—Duality and Sensitivity Analysis

6.1 Duality

Associated with each linear program, which we shall call the *primal*, is another linear program called its *dual*.

Let the primal problem be:
Maximize

$$z_x = c_1 x_1 + c_2 x_2 + \ldots + c_n x_n$$

subject to

$$a_{11} x_1 + a_{12} x_2 + \ldots + a_{1n} x_n \leqslant b_1,$$
$$a_{21} x_1 + a_{22} x_2 + \ldots + a_{2n} x_n \leqslant b_2, \qquad (6.1.1)$$
$$\vdots$$
$$a_{m1} x_1 + a_{m2} x_2 + \ldots + a_{mn} x_n \leqslant b_m,$$
$$x_j \geqslant 0, \qquad j = 1, 2, \ldots, n.$$

Its dual is:
Minimize

$$z_y = b_1 y_1 + b_2 y_2 + \ldots + b_m y_m$$

subject to

$$a_{11} y_1 + a_{21} y_2 + \ldots + a_{m1} y_m \geqslant c_1,$$
$$a_{12} y_1 + a_{22} y_2 + \ldots + a_{m2} y_m \geqslant c_2, \qquad (6.1.2)$$
$$\vdots$$
$$a_{1n} y_1 + a_{2n} y_2 + \ldots + a_{mn} y_m \geqslant c_n,$$
$$y_i \geqslant 0, \qquad i = 1, 2 \ldots m.$$

The relationship between these two problems is more apparent if we use matrix notation. Let

$$A = \begin{bmatrix} a_{11} & a_{12} & \cdots & a_{1n} \\ a_{21} & a_{22} & \cdots & a_{2n} \\ \vdots & & & \\ a_{m1} & a_{m2} & \cdots & a_{mn} \end{bmatrix}, \quad A^T = \begin{bmatrix} a_{11} & a_{21} & \cdots & a_{m1} \\ a_{12} & a_{22} & \cdots & a_{m2} \\ \vdots & & & \\ a_{1n} & a_{2n} & \cdots & a_{mn} \end{bmatrix}$$

be the matrices of coefficients of (6.1.1) and (6.1.2) respectively, and let vectors be written as column vectors and let row vectors be written as transposes of column vectors. Then (6.1.1) and (6.1.2) may be written respectively:

$$\text{Maximize } z_x = c^T x \quad \text{subject to } Ax \le b, \qquad x \ge 0;$$

$$\text{Minimize } z_y = b^T y \quad \text{subject to } A^T y \ge (c^T)^T = c, \qquad y \ge 0$$

where

$$x = \begin{bmatrix} x_1 \\ x_2 \\ \vdots \\ x_n \end{bmatrix}, \quad y = \begin{bmatrix} y_1 \\ y_2 \\ \vdots \\ y_m \end{bmatrix}, \quad b = \begin{bmatrix} b_1 \\ b_2 \\ \vdots \\ b_m \end{bmatrix}, \quad c = \begin{bmatrix} c_1 \\ c_2 \\ \vdots \\ c_n \end{bmatrix}$$

are column vectors. If we take (6.1.2), treat it as the primal problem and write

$$\text{Maximize } z_x = c^T x \quad \text{subject to } (A^T)^T x \le (b^T)^T = b, \qquad x \ge 0,$$

we see that we are back to (6.1.1). Thus the dual of the dual is the primal, and it is indifferent which one of the problems is called the primal.

This duality also exists if (6.1.1) is modified so that some constraint, say the ith, is in the form of an equation instead of an inequality. The dual problem will then have a corresponding modification with the ith dual variable y_i unrestricted in sign instead of nonnegative. Similarly, if some x_j in the primal is unrestricted in sign, then the corresponding jth constraint in the dual problem is an equation instead of an inequality. This symmetry may be worked out by the reader after recalling from Section 5.1 that an unrestricted variable, say, x_j may be replaced by $x_j' - x_j''$, and that an equation is equivalent to two inequalities with $\le$ for one and $\ge$ for the other.

As illustration let the primal problem be (5.2.9), which is repeated below for easier reference:

Maximize

$$z_x = x_1 + 3x_2 + 4x_3$$

subject to

$$3x_1 + 2x_2 \qquad \le 13, \tag{6.1.3}$$

$$x_2 + 3x_3 \le 17,$$

$$2x_1 + x_2 + x_3 \le 13,$$

$$x_j \ge 0, \quad j = 1, 2, 3.$$

Its dual is:
Minimize

$$z_y = 13y_1 + 17y_2 + 13y_3$$

subject to

$$3y_1 \qquad + 2y_3 \geqslant 1, \qquad\qquad (6.1.4)$$
$$2y_1 + y_2 + y_3 \geqslant 3,$$
$$3y_2 + y_3 \geqslant 4,$$
$$y_i \geqslant 0, \qquad i = 1, 2, 3.$$

Suppose that x_3 is unrestricted in sign. (6.1.3) now becomes:
Maximize

$$z_x = x_1 + 3x_2 + 4x_3' - 4x_3''$$

subject to

$$3x_1 + 2x_2 \qquad\qquad\qquad \leqslant 13, \qquad\qquad (6.1.5)$$
$$x_2 + 3x_3' - 3x_3'' \leqslant 17,$$
$$2x_1 + x_2 + x_3' - x_3'' \leqslant 13,$$
$$x_1 \geqslant 0, \qquad x_2 \geqslant 0, \qquad x_3' \geqslant 0, \qquad x_3'' \geqslant 0;$$

and (6.1.4) becomes:
Minimize

$$z_y = 13y_1 + 17y_2 + 13y_3$$

subject to

$$3y_1 \qquad + 2y_3 \geqslant 1,$$
$$2y_1 + y_2 + y_3 \geqslant 3, \qquad\qquad (6.1.6)$$
$$3y_2 + y_3 \geqslant 4,$$
$$-3y_2 - y_3 \geqslant -4,$$
$$y_i \geqslant 0, \qquad i = 1, 2, 3.$$

The last two inequalities of (6.1.6) are equivalent to the equation $3y_2 + y_3 = 4$.

The intimate relationship between the primal and the dual may be seen from the following theorems. We shall state the theorems one after another and leave the longer proofs to the next section so that the reader can more easily see the close relationship without diverting attention to the details of a proof. In what follows, we shall use (6.1.1) as the primal problem but keep in mind the symmetry between the primal and the dual. First, let us write (6.1.1) and (6.1.2) respectively as:

(0) $z_x - c_1 x_1 - c_2 x_2 - \ldots - c_n x_n$ $\qquad\qquad\qquad = 0;$

(1) $a_{11} x_1 + a_{12} x_2 \qquad + a_{1n} x_n + x_{n+1} \qquad\qquad\qquad = b_1;$

(2) $a_{21} x_1 + a_{22} x_2 \qquad + a_{2n} x_n \qquad\quad + x_{n+2} \qquad\qquad = b_2;$ $\qquad$ (6.1.7)

$\vdots$

(m) $a_{m1} x_1 + a_{m2} x_2 \qquad + a_{mn} x_n \qquad\qquad\qquad + x_{n+m} = b_m;$

and

(0) $z_y - b_1 y_1 - b_2 y_2 - \ldots - b_m y_m$ $\qquad\qquad\qquad\qquad = 0;$

(1) $a_{11} y_1 + a_{21} y_2 \qquad + a_{m1} y_m - y_{m+1} \qquad\qquad\quad = c_1;$

(2) $a_{12} y_1 + a_{22} y_2 \qquad + a_{m2} y_m \qquad\quad - y_{m+2} \qquad = c_2;$ $\qquad$ (6.1.8)

$\vdots$

(n) $a_{1n} y_1 + a_{2n} y_2 \qquad + a_{mn} y_m \qquad\qquad\qquad - y_{m+n} = c_n.$

We shall use the superscript "o" to denote the optimal value of a function or variable. Thus, z_x^o is the optimal value of z_x and x_j^o is the optimal value of x_j. Also, let s_j^o be *simplex multipliers* for the primal problem. That is, s_j^o is the amount by which the original coefficient $-c_j$ of x_j in the initial objective equation (0) has been changed in the process of carrying out the simplex method; therefore the coefficient of x_j in the final equation (0) is $s_j^o - c_j$, for $j = 1, 2, \ldots, n + m$. Notice that $c_j = 0$ for $j = n + 1, \ldots, n + m$. Hence s_{n+i}^o (for $i = 1, \ldots, m$) is simply the coefficient of the $(n + i)$th variable in the final objective equation (0). That is, the final equation (0) for the primal is:

$$z_x^o = z_x + (s_1^o - c_1)x_1 + (s_2^o - c_2)x_2 + \ldots$$
$$+ (s_n^o - c_n)x_n + s_{n+1}^o x_{n+1} + \ldots + s_{n+m}^o x_{n+m}.$$

Similar theorems may be stated for the dual, but we shall confine ourselves to the primal problem and let the reader work out the dual counterparts.

In the initial set of equations given by (6.1.7), the coefficient of x_{n+i} is 1 in equation (i) and 0 in all the other equations, and the values of x_j for $j = 1, 2, \ldots, n$ are all 0. Therefore s_{n+i}^o times the initial equation (i) must have been eventually added (directly or indirectly) through the simplex process to the initial equation (0). This leads directly to:

Theorem 6.1.1. $z_x^o = \sum_{i=1}^m s_{n+i}^o b_i$ and $s_j^o = \sum_{i=1}^m s_{n+i}^o a_{ij}$, for $j = 1, 2, \ldots, n$. $\qquad\square$

For our example of Table 5.3.1, by way of illustration,

$$s_4^o = \tfrac{5}{6}, \qquad s_5^o = \tfrac{4}{3}, \qquad s_6^o = 0,$$

$$s_1^o = \tfrac{5}{2} = \left(\tfrac{5}{6}\right)(3) + \left(\tfrac{4}{3}\right)(0) + (0)(2).$$

In equation (0), when the coefficient of $x_5 = x_{n+2}$ is increased from 0 to

$4/3$, z_x is increased from 0 to $(4/3)17 = s_5^o \cdot b_2$. When, at the next iteration, the coefficient of $x_4 = x_{n+1}$ is increased from 0 to $5/6$, z_x is increased from $(4/3)17$ to $(4/3)(17) + (5/6)(13) = s_5^o \cdot b_2 + s_4^o \cdot b_1$.

Theorem 6.1.2. *For any feasible solutions of the primal and dual problems,* $z_x \leqslant z_y$. $\square$

From Theorem 6.1.2 we see that if one problem has an unbounded optimal solution, then the other cannot have a feasible solution. If that other problem had a feasible solution, it would act as a bound for any feasible (optimal or nonoptimal) solution of the first.

Theorem 6.1.3. *Suppose the primal has an optimal solution. Then the dual also has an optimal solution and the optimal value of the ith dual variable is the same as the coefficient of the ith slack variable of the primal in the final equation* (0). *That is,*

$$y_i^o = s_{n+i}^o \quad \text{for } i = 1, 2, \ldots, m.$$ $\square$

The next theorem is an immediate consequence of Theorems 6.1.1 and 6.1.3:

Theorem 6.1.4. $s_j^o - c_j = \sum_{i=1}^{m} y_i^o a_{ij} - c_j$ *for* $j = 1, 2, \ldots, n$. *The coefficient of* x_j *in the final equation* (0) *is the same as the difference between the left- and right-hand sides of the jth constraint of the dual using the optimal dual solution.* $\square$

But, by definition, $y_{m+j}^o = \sum_{i=1}^{m} a_{ij} y_i^o - c_j$ for $j = 1, 2, \ldots, n$. Therefore we have:

Theorem 6.1.5. $y_{m+j}^o = s_j^o - c_j$ *for* $j = 1, 2, \ldots, n$. *The optimal value of the jth dual slack variable is the same as the coefficient of the jth variable in the final primal equation* (0). $\square$

The next theorem is a fundamental one:

Theorem 6.1.6 (a) (Dual or Duality theorem). *Suppose that both the primal and dual problems have finite feasible solutions. Then there exist finite optimal solutions* $\mathbf{x}^o$ *and* $\mathbf{y}^o$ *for both problems yielding the same optimal value for their objective functions, i.e.,*

$$z_x^o = z_y^o.$$ $\square$

A slightly stronger version of the Dual theorem reads:

Theorem 6.1.6 (b). *If a feasible solution exists for one of the two* (primal or dual) *problems and yields a finite optimal objective function value, then a*

*feasible optimal solution exists for the other problem yielding the same
optimal objective function value.*   ☐

We may summarize the above two statements and state:

*A necessary and sufficient condition for one (hence both) of the problems to
have finite optimal solutions is that both have feasible solutions.* Therefore
the simplex method may be thought of as a process to obtain a feasible
solution to the dual while maintaining feasibility for the primal. Optimal
solutions are obtained as soon as feasible solutions for both are reached.

Theorems 6.1.3 and 6.1.5 lead to:

Theorem 6.1.7 (Complementary slackness theorem).

$$y_i^o = 0 \quad \text{if} \quad x_{n+i}^o > 0 \quad \text{for } i = 1, 2, \ldots, m,$$

and

$$y_{m+j}^o = 0 \quad \text{if} \quad x_j^o > 0 \quad \text{for } j = 1, 2, \ldots, n.$$

Proof. If $x_k^o > 0$ (for $k = 1, 2, \ldots, n, \ldots, n+m$), it must be a basic variable.
Hence its coefficient in the final objective equation (0) must be 0, and this
coefficient is equal to the optimal value of the corresponding dual variable.
☐

The complementary slackness theorem may be restated as

$$y_i^o \cdot x_{n+i}^o = 0,$$
$$y_{m+j}^o \cdot x_j^o = 0.$$

Or, since $x_{n+i} = b_i - \Sigma_j a_{ij} x_j$ and $y_{m+j} = \Sigma_i a_{ij} y_i - c_j$ by definition,

$$y_i^o \cdot \left(\sum_{j=1}^{n} a_{ij} x_j^o - b_i \right) = 0,$$

$$x_j^o \cdot \left(\sum_{i=1}^{m} a_{ij} y_i^o - c_j \right) = 0.$$

This means that if a constraint in the final set of equations, say the ith in
the primal (or jth in the dual), is a strict inequality so that there is a slack
variable x_{n+i}^o (or y_{m+j}^o) in that constraint, then the corresponding ith
variable in the dual (or jth variable in the primal) problem is equal to 0.
And if the ith variable in the dual (or jth variable in the primal) is positive,
then the ith constraint in the primal (or jth constraint in the dual) must be
an equality.

A stronger version of the complementary slackness theorem is:

*If x_j^F, for $j = 1, \ldots, n$, and y_i^F, for $i = 1, \ldots, m$ are feasible solutions for the
primal and dual problems respectively, then they are optimal solutions if and*

only if

$$y_i^F \cdot \left(\sum_{j=1}^{n} a_{ij} x_j^F - b_i \right) = 0 \quad for \ i = 1, \ldots, m$$

and

$$x_j^F \cdot \left(\sum_{i=1}^{m} a_{ij} y_i^F - c_j \right) = 0 \quad for \ j = 1, \ldots, n.$$

The following theorem shows the relationship between the objective functions of the two problems even before optimality is achieved.

Theorem 6.1.8. *Assume that finite feasible (hence optimal) solutions exist for both the primal and dual problems. At an iteration of the simplex method, let* $\mathbf{x}^F = (x_1^F, x_2^F, \ldots, x_{n+m}^F)^T$ *be a basic feasible nonoptimal solution for the primal and let* $s_j^F - c_j$ *be the coefficient of* x_j^F *in the current equation* (0) *for* $j = 1, 2, \ldots, n+m$. *(Remember that* $c_j = 0$ *for* $j = n+1, \ldots, n+m$.) *Let* $\mathbf{y}' = (y_1', y_2', \ldots, y_{m+n}')^T$ *be a basic but infeasible solution for the dual, where* $y_i' = s_{n+i}^F$ *for* $i = 1, \ldots, m$, *and* $y_{m+j}' = s_j^F - c_j$ *for* $j = 1, \ldots, n$. *Then*

$$z_x^F = z_y', \quad where \ z_x^F = \sum_{j=1}^{n} c_j x_j^F \quad and \quad z_y' = \sum_{i=1}^{m} b_i y_i'.$$

Proof. By definition, $z_x = \sum_{j=1}^{n} c_j x_j$, and this remains true for the specific values x_j^F, so that $z_x^F = \sum_{j=1}^{n} c_j x_j^F$. Similarly, $z_y' = \sum_{i=1}^{m} b_i y_i'$ by definition. The same arguments that led to Theorem 6.1.1, with "o" replaced by "F", lead to the conclusion that

$$z_x^F = \sum_{i=1}^{m} s_{n+i}^F \cdot b_i = \sum_{i=1}^{m} y_i' \cdot b_i = z_y'. \qquad \square$$

Some of the relationships between the two problems may be summarized in the table below. We consider the most general case in which the primal problem consists of k equalities, r "$\leqslant$" conditions, s "$\geqslant$" conditions, p variables are constrained to be nonnegative, and $n-p$ variables are unconstrained in sign. The letters A, B, C, D, E, and F are submatrices of coefficients. Before going to the dual problem, we first multiply constraints $k+r+1, k+r+2, \ldots, k+r+s$ by -1 so as to reverse the inequalities from "$\geqslant$" to "$\leqslant$". Then we obtain the chart for the dual, where the superscript T stands for the transpose of the original submatrix.

Let us illustrate some of the relationships by looking at our example given in (6.1.3) and (6.1.4). In view of the complementarity of the solutions of the two problems, we can actually solve both the primal and the dual together using the same tableau. For the primal problem, the nonbasic variables are placed at the top and the objective function z_x and basic variables on the left-hand side. The signs of the coefficients are reversed from those of the simplex tableau of Table 5.3.1 to accommodate the dual problem. For the dual problem, the nonbasic variables y_1, y_2, y_3 (which correspond to the basic variables x_4, x_5, x_6 of the primal in the initial

Table 6.1.1.

$$\text{PRIMAL} \qquad \max z_x = \sum_{j=1}^{n} c_j x_j$$

	Nonnegative	Unrestricted		
	$x_1 \ldots x_p$	$x_{p+1} \ldots x_n$		
$\begin{matrix}1\\ \vdots\\ k\end{matrix}$	A	B	$=$	$\begin{matrix}b_1\\ \vdots\\ b_k\end{matrix}$
$\begin{matrix}k+1\\ \vdots\\ k+r\end{matrix}$	C	D	$\leqslant$	$\begin{matrix}b_{k+1}\\ \vdots\\ b_{k+r}\end{matrix}$
$\begin{matrix}k+r+1\\ \vdots\\ k+r+s\\ =m\end{matrix}$	E	F	$\geqslant$	$\begin{matrix}b_{k+r+1}\\ \vdots\\ b_m\end{matrix}$
	$c_1 \ldots c_p$	$c_{p+1} \ldots c_n$		

↓

Feasible but nonoptimal primal solutions

↓

$$z_x^o = z_y^o$$

↑

Infeasible but "better than optimal" dual solutions

↑

	Unrestricted	Nonnegative			
	$y_1 \ldots y_k$	$y_{k+1} \ldots y_{k+r}$	$y_{k+r+1} \ldots y_m$		
$\begin{matrix}1\\ \vdots\\ p\end{matrix}$	A^T	C^T	$-E^T$	$=$	$\begin{matrix}c_1\\ \vdots\\ c_p\end{matrix}$
$\begin{matrix}p+1\\ \vdots\\ n\end{matrix}$	B^T	D^T	$-F^T$	$\geqslant$	$\begin{matrix}c_{p+1}\\ \vdots\\ c_n\end{matrix}$
	$b_1 \ldots b_k$	$b_{k+1} \ldots b_{k+r}$	$-b_{k+r+1} \ldots -b_m$		

$$\text{DUAL} \qquad \min z_y = \sum_{i=1}^{m} b_i y_i$$

Table 6.1.2.

max		x_1	x_2	x_3	
$z_x =$	0	c_1	c_2	c_3	
$x_4 =$	b_1	$-a_{11}$	$-a_{12}$	$-a_{13}$	y_1
$x_5 =$	b_2	$-a_{21}$	$-a_{22}$	$-a_{23}$	y_2
$x_6 =$	b_3	$-a_{31}$	$-a_{32}$	$-a_{33}$	y_3
	$z_y =$	$-y_4 =$	$-y_5 =$	$-y_6 =$	min

iteration), are placed on the right-hand side opposite their corresponding complementary variables and, similarly, the objective function z_y and the basic variables y_4, y_5, y_6 (with a negative sign because the inequalities for the dual are $\geqslant$) are placed at the bottom, as shown in Table 6.1.2.

From Table 6.1.2, for example, we read:
Maximize

$$z_x = 0 + c_1 x_1 + c_2 x_2 + c_3 x_3,$$

$$x_4 = b_1 - a_{11} x_1 - a_{12} x_2 - a_{13} x_3,$$

$$\vdots \qquad \qquad ;$$

Minimize

$$z_y = 0 + b_1 y_1 + b_2 y_2 + b_3 y_3,$$

$$-y_4 = c_1 - a_{11} y_1 - a_{21} y_2 - a_{31} y_3,$$

$$\vdots$$

$$-y_6 = c_3 - a_{13} y_1 - a_{23} y_2 - a_{33} y_3.$$

The combined initial tableau for (6.1.3) and (6.1.4) appears in Table 6.1.3. Notice that in this initial iteration the basic solution for the primal $(0,0,0,13,17,13)$ is feasible but not optimal. The basic solution for the dual $(0,0,0,-1,-3,-4)$ is "better than optimal" but not feasible. $s_j^F = 0$ for $j = 1, \ldots, 6$. Thus $y_1' = y_2' = y_3' = 0$, $y_4' = -c_1 = -1$, $y_5' = -c_2 = -3$, $y_6' = -c_3 = -4$.
The combined tableau for the second iteration appears in Table 6.1.4.

Table 6.1.3.

max		x_1	x_2	x_3	
$z_x =$	0	1	3	4	
$x_4 =$	13	-3	-2	0	y_1
$x_5 =$	17	0	-1	-3	y_2
$x_6 =$	13	-2	-1	-1	y_3
	$z_y =$	$-y_4 =$	$-y_5 =$	$-y_6 =$	min

Table 6.1.4.

		x_1	x_2	x_5	
$z_x =$	$\frac{68}{3}$	1	$\frac{5}{3}$	$-\frac{4}{3}$	
$x_4 =$	13	-3	-2	0	y_1
$x_3 =$	$\frac{17}{3}$	0	$-\frac{1}{3}$	$-\frac{1}{3}$	y_6
$x_6 =$	$\frac{22}{3}$	-2	$-\frac{2}{3}$	$\frac{1}{3}$	y_3
	$z_y =$	$-y_4 =$	$-y_5 =$	$-y_2 =$	

In this iteration, x_3 has replaced x_5 as the basic variable in the primal, and correspondingly, y_6 replaces y_2 as the nonbasic variable in the dual. The basic solution for the primal is $(0, 0, 17/3, 13, 0, 22/3)$, which is still feasible but not optimal. Equation (0) for the primal is $z_x - 1 \cdot x_1 - (5/3)x_2 + (4/3)x_5 = 68/3$. Hence the complementary dual solution is $(0, 4/3, 0, -1, -5/3, 0)$, which is still "better than optimal" but not feasible. (Notice that $y_1 = 0 = s_4$, $y_2 = 4/3 = s_5$, $y_3 = 0 = s_6$, $y_4 = -1 = s_1 - c_1$, $y_5 = -5/3 = s_2 - c_2$, $y_6 = 0 = s_3 - c_3$.) $z_x = 1 \cdot 0 + 3 \cdot 0 + 4 \cdot (17/3) = 68/3 = 0 \cdot 13 + (\frac{4}{3}) \cdot 17 + 0 \cdot 13 = z_y$, as stated in Theorem 6.1.8.

The combined tableau for the final iteration appears in Table 6.1.5.

In this iteration, x_2 has replaced x_4 as the basic variable in the primal, and y_5 replaces y_1 as the nonbasic variable in the dual. The basic solution for the primal is $(0, 13/2, 7/2, 0, 0, 3)$, which is feasible and optimal, and equation (0) is $z_x^o + (3/2)x_1 + (5/6)x_4 + (4/3)x_5 = 67/2$. The corresponding dual solution is $(5/6, 4/3, 0, 3/2, 0, 0)$, which is now feasible, hence optimal, with $z_x^o = 67/2 = z_y^o$ (Theorem 6.1.6). The optimal values of y_1, y_2, and y_3 are the coefficients of x_4, x_5, and x_6 and the optimal values of y_4, y_5, and y_6 are the coefficients of x_1, x_2, and x_3 in equation (0) of the primal (Theorems 6.1.3 and 6.1.5), and $z_x^o = (5/6) \cdot 13 + (4/3) \cdot 17 + 0 \cdot 13$ (Theorem 6.1.1). Moreover, (Theorem 6.1.4), the coefficients of x_1^o, x_2^o, and x_3^o are, respectively,

$$\tfrac{3}{2} = 3 \cdot \tfrac{5}{6} + 0 \cdot \tfrac{4}{3} + 2 \cdot 0 - 1,$$

$$0 = 2 \cdot \tfrac{5}{6} + 1 \cdot \tfrac{4}{3} + 1 \cdot 0 - 3,$$

$$0 = 0 \cdot \tfrac{5}{6} + 3 \cdot \tfrac{4}{3} + 1 \cdot 0 - 4.$$

Finally, (Theorem 6.1.7), $y_i^o \cdot x_{n+i}^o = 0$ as seen in $(5/6) \cdot 0$, $(4/3) \cdot 0$, and $0 \cdot 3$,

Table 6.1.5.

		x_1	x_4	x_5	
$z_x^o =$	$\frac{67}{2}$	$-\frac{3}{2}$	$-\frac{5}{6}$	$-\frac{4}{3}$	
$x_2 =$	$\frac{13}{2}$	$-\frac{3}{2}$	$-\frac{1}{2}$	0	y_5
$x_3 =$	$\frac{7}{2}$	$\frac{1}{2}$	$\frac{1}{6}$	$-\frac{1}{3}$	y_6
$x_6 =$	3	-1	$\frac{1}{3}$	$\frac{1}{3}$	y_3
	$z_y^o =$	$-y_4 =$	$-y_1 =$	$-y_2 =$	

and $y_{m+j}^{\circ}\cdot x_j^{\circ}=0$ as seen in $(3/2)\cdot 0$, $0\cdot(13/2)$, $0\cdot(7/2)$. The reader, by the way, may have noticed the same solution in Section 5.5.

The combined tableau also brings out clearly the complementary nature of the feasibility and optimality conditions of the primal and dual problems. For the primal, the feasibility condition is that the first column of the combined tableau (column of b's) should be all nonnegative. Since the dual problem can be thought of as lying on its side in the combined tableau, the first column is the column of coefficients of the dual variables, and the optimality condition of the dual problem is that these coefficients should be nonnegative. Therefore, the feasible condition of the primal corresponds to the optimal condition of the dual. Next, except for the upper left corner, the first row of the combined tableau represents the coefficients of the primal variables, and, at the same time, the values of the basic dual variables with their signs reversed $(-y_i)$. The optimality condition of the primal is that this row should be nonpositive, which corresponds to the feasibility condition of the dual. Therefore, when feasibility is obtained in both, optimality will also have been obtained in both (Theorem 6.1.6). The reader is again reminded that the combined tableau (Table 6.1.2) differs from the single simplex tableau (Table 5.3.1) in that in the combined tableau equation (0) is written as

$$z_x = 0 + c_1 x_1 + c_2 x_2 + \dots + c_n x_n,$$

whereas in the single simplex tableau, equation (0) is written as

$$z_x - c_1 x_1 - c_2 x_2 - \dots - c_n x_n = 0.$$

In this paragraph, the words "nonpositive" and "nonnegative" are used in conjunction with the combined tableau. The signs should be reversed if we refer to the single tableau.

6.2 Proofs of some theorems

We shall take a brief look at the proofs of some of the theorems which we postponed to this section.

PROOF OF THEOREM 6.1.2. We recall from (6.1.1) and (6.1.2) that if $\mathbf{x}=(x_1,\dots,x_n)^{\mathrm{T}}$ and $\mathbf{y}=(y_1,\dots,y_m)^{\mathrm{T}}$ are feasible solutions, then

$$\sum_{j=1}^{n} a_{ij} x_j \leqslant b_i, \qquad \sum_{i=1}^{m} a_{ij} y_i \geqslant c_j, \qquad x_j \geqslant 0, y_i \geqslant 0,$$

for $j=1,\dots,n$, $i=1,\dots,m$. Now

$$z_x = \sum_{j=1}^{n} c_j x_j \leqslant \sum_{j=1}^{n} \left(\sum_{j=1}^{m} a_{ij} y_i \right) x_j$$

$$= \sum_{i=1}^{m} y_i \left(\sum_{j=1}^{n} a_{ij} x_j \right)$$

$$\leqslant \sum_{i=1}^{m} y_i b_i = z_y \qquad \qquad \square$$

PROOF OF THEOREM 6.1.3. Let $y_1^* = s_{n+i}^o$ for $i = 1, \ldots, m$. We wish to show that $(y_1^*, \ldots, y_m^*)$ is a feasible solution for the dual problem and that it is also optimal.

The optimality condition of the primal is that all coefficients in the final equation (0) are nonnegative. Hence $s_{n+i}^o \geqslant 0$, for $i = 1, \ldots, m$, thus fulfilling the nonnegativity requirement for the dual variables y_i^*.

Moreover, the same optimality condition of the primal implies that

$$s_j^o - c_j \geqslant 0 \quad \text{for } j = 1, \ldots, n.$$

Combining this with Theorem 6.1.1, we have

$$\sum_{i=1}^m y_i^* a_{ij} = \sum_{i=1}^m s_{n+i}^o a_{ij} = s_j^o \geqslant c_j \quad \text{for } j = 1, \ldots, n.$$

Therefore $(y_1^*, \ldots, y_m^*)$ is a feasible solution. To show that it is also optimal, we know from Theorem 6.1.2 that

$$z_x^o \leqslant \sum_{i=1} b_i y_i$$

for any feasible solution $(y_1, \ldots, y_m)$ of the dual. But, from Theorem 6.1.1, $z_x^o = \sum_{i=1}^m b_i s_{n+i}^o$. Hence $\sum_{i=1}^m b_i s_{n+i}^o = \sum_{i=1}^m b_i y_i^*$ must be the smallest value of the objective function z_y of the dual, thus establishing that $(y_1^*, \ldots, y_m^*)$ is indeed optimal. $\square$

PROOF OF THEOREM 6.1.6. It is assumed that both the primal and dual problems have finite feasible solutions. As we commented in Section 6.1, Theorem 6.1.2 then implies both must have finite optimal solutions (since each will be bounded, the primal from above and the dual from below). Theorem 6.1.1 has established that

$$z_x^o = \sum_{i=1}^m b_i s_{n+i}^o$$

and Theorem 6.1.3 says that

$$z_y^o = \sum b_i s_{n+i}^o.$$

Therefore $z_x^o = z_y^o$. $\square$

PROOF OF THEOREM 6.1.7 (Complementary slackness theorem). We have already proved that if $(x_1^o, \ldots, x_n^o)$ and $(y_1^o, \ldots, y_m^o)$ are optimal solutions, then

$$y_i^o \cdot \left(\sum_j a_{ij} x_j^o - b_i \right) = 0 \quad \text{for } i = 1, \ldots, m,$$

$$x_j^o \cdot \left(\sum_i a_{ij} y_i^o - c_j \right) = 0 \quad \text{for } j = 1, \ldots, n.$$

Suppose x_j^F, for $j = 1, \ldots, n$, and y_i^F, for $i = 1, \ldots, m$, are feasible solutions such that

$$y_i^F \cdot \left(\sum_j a_{ij} x_j^F - b_i \right) = 0 \quad \text{for } i = 1, \dots, m,$$

$$x_j^F \cdot \left(\sum_i a_{ij} y_i^F - c_j \right) = 0 \quad \text{for } j = 1, \dots, n.$$

We wish to show that they are optimal.

$$\sum_i y_i^F \cdot \left(\sum_j a_{ij} x_j^F - b_i \right) = \sum_j x_j^F \cdot \left(\sum_i a_{ij} y_i^F - c_j \right) = 0$$

implies that

$$z_y = \sum_i y_i^F b_i = \sum_j x_j^F c_j = z_x.$$

It follows from Theorems 6.1.2 and 6.1.6 that $(x_1^F, \dots, x_n^F)$ and $(y_1^F, \dots, y_m^F)$ are optimal solutions. □

6.3 Economic interpretation of duality

One of the main applications of linear programming is in economics. It is a topic too big to be covered in this volume. However, the role of duality in economic analysis is particularly interesting.

Our primal program may be a problem in production for a firm which produces n different commodities subject to m capacity or scarce input constraints. Management wishes to determine x_j, the amount of commodity j to be produced, so as to maximize total profit z_x. If c_j is the unit profit of commodity j, then

$$z_x = \sum_{j=1}^n c_j x_j.$$

Let:

$a_{ij} =$ the number of units of scarce resource (input) i used to produce commodity j;

$b_i =$ the amount of resource i available to the firm.

Thus $a_{i1}x_1 + a_{i2}x_2 + \dots + a_{in}x_n \leqslant b_i$, for $i = 1, \dots, m$, is the requirement that the total amount of resource i used in the production of the n commodities cannot exceed the capacity limitations.

Each dual variable y_i may be thought of as the unit accounting value of resource i. Management imputes all of the firm's profits to its scarce resources; that is, the y_i should be so chosen that the sum of the unit accounting values of the m resources going into the production of one unit of commodity j is high enough to account for the unit profit from commodity j. Since $a_{1j}y_1 + a_{2j}y_2 + \dots + a_{mj}y_m$ is this sum, we have

$$a_{1j}y_1 + a_{2j}y_2 + \dots + a_{mj}y_m \geqslant c_j \quad \text{for } j = 1, \dots, n.$$

In view of this requirement of no accounting profit, the reader may naturally wonder why management does not assign arbitrarily high accounting values to its m scarce resources. This takes us to the dual objective function $z_y = b_1 y_1 + \ldots + b_m y_m$, which represents the total value of all the inputs available to the firm. The dual problem is that the firm seeks the smallest valuation of the firm's total inputs which completely accounts for the profits from all its outputs; that is, the prices of the resources should be such as to minimize their total cost to the firm.

The dual slack variable y_{m+j}, for $j = 1, \ldots, n$, is

$$y_{m+j} = (a_{1j} y_1 + a_{2j} y_2 + \ldots + a_{mj} y_j) - c_j.$$

Therefore $y_{m+j} =$ (the accounting value of the inputs used in producing one unit of output j) $-$ (the profit per unit of output j), and may be thought of as an accounting-loss figure for commodity j. If y_{m+j} is positive, it means that the resources used to produce commodity j are worth more than the profit yielded by that commodity.

The primal slack variable x_{n+i}, for $i = 1, \ldots, m$, is

$$x_{n+i} = b_i - (a_{i1} x_1 + a_{i2} x_2 + \ldots + a_{in} x_n).$$

That is, $x_{n+i} =$ (the amount of resource i available) $-$ (the amount of resource i used to produce the n commodities) and represents the unused capacity of resource i. Let us summarize the relationships as follows.

	Primal variables (physical quantities)	Dual variables (monetary magnitudes)
Output	x_j	y_{m+j}
Input	x_{n+i}	y_i

Notice that the primal variables represent physical quantities. For instance, if the three outputs of our example of (6.1.3) and (6.1.4) are tractors, motorcycles, and snowmobiles, and the three resources are warehouse space, machine time, and labor, then x_1, x_2, and x_3 would be measured in units of hundreds or thousands, and x_4, x_5, and x_6 in square feet, hours, and person-weeks. The dual variables, on the other hand, would be monetary units such as dollars.

Theorem 6.1.2 now tells us that the total value imputed to inputs should be no less than the contribution of the inputs to total profit, i.e., $z_y \geqslant z_x$. The dual theorem further says that in an optimum situation the total accounting value assigned to inputs should not exceed the total net profit from the sale of outputs, and that, in fact, they are equal. Total accounting value of inputs can be viewed as an opportunity cost. The Dual theorem says profits are equal to the opportunity cost of inputs. Theorem 6.1.7 then says that in the optimum situation, the firm will produce only those commodities j whose accounting-loss figures y_{m+j} are zero, i.e., $x_j^o \cdot y_{m+j}^o = 0$, for $j = 1, \ldots, n$. Also, only those inputs i which are used to capacity (hence $x_{n+i}^o = 0$) will be assigned a positive ($y_i^o > 0$) accounting valuation, i.e., $y_i^o \cdot x_{n+i}^o = 0$, for $i = 1, \ldots, m$. This implies that the price of a resource which is oversupplied (a sort of "free good") is zero by the law of supply and demand. Thus inputs in excess supply are given a zero accounting value and commodities which are associated with a positive accounting loss should not, optimally, be produced. By looking at the optimal solution of the dual problem we can tell which outputs should not be produced and which inputs will not be used to capacity. In our example, since $y_4^o = \frac{3}{2}$, commodity 1 should not be produced. Since $y_3^o = 0$, resource 3 will not be used to capacity.

Moreover, it can be proved that

$$y_i^o = \frac{\partial z_x^o}{\partial b_i}$$

if the derivative exists. That is, y_i^o is equal to the marginal contribution of input i to profit. It is the rate at which profit will increase (decrease) if b_i, the amount of resource i, is increased (decreased) in such a way that y_i^o is still equal to s_{n+i}^o. Without giving a proof, let us indicate the reasoning as follows. Recall from Theorem 6.1.1 that s_{n+i}^o times the original equation (i) is added directly or indirectly to the initial equation (0) to obtain the final equation (0). If b_i is increased (decreased) by an amount k, then z_x^o will be increased (decreased) by $k \cdot s_{n+i}^o$, assuming that we have the same optimal basis as before. Since $y_i^o = s_{n+i}^o$, y_i^o is this marginal contribution to z_x^o as claimed.

The economic interpretation of y_{m+j}^o, the accounting-loss figure, is equally interesting. It is the opportunity cost of commodity j. A positive accounting-loss figure, such as y_4^o $(= \frac{3}{2})$ of our example, compared with zero loss figure for y_5^o and y_6^o, means that although commodity 1 yields a profit of \$100 or \$1000 $(c_1 = 1)$, the firm still suffers a loss of \$150 or \$1500 $(y_4^o = \frac{3}{2})$ by transferring resources from the production of commodities 2 and 3 to the production of commodity 1.

We may summarize the situation by concluding that in the primal problem management tries to maximize profit directly by determining the level of outputs of the n commodities subject to the m capacity limitations. In the dual, management attacks the same problem indirectly by allocating scarce resources among the production of the n commodities until the minimum total value (or opportunity cost) of the resources is equal to the maximum total profit that can be obtained through their use.

Finally, it has been suggested that duality theory aids in decentralized planning by the use of dual pricing. For a large firm with several plants scattered in several locations, top management can devise a master plan and allocate its scarce resources among the various plants. If, instead, central management can somehow calculate the values of the dual accounting prices y_i, then it is possible for each plant manager to take as much input as he wishes as long as the items produced by that plant will yield no accounting loss $(y_{m+j}^o = 0)$. A good plant manager will then decide to take only those inputs which will be used to capacity and produce only those items which will yield unit profits high enough to cover the accounting-value charge on the inputs used to produce them.

6.4 Dual simplex method

When we solved the dual problem simultaneously with the primal problem in Section 6.1, we in effect started with a "better than optimal" but infeasible solution and worked towards feasibility. As we saw in Chapter 5, when an initial basic feasible solution using slack variables is not available, it may be necessary to introduce several artificial variables to construct an artificial initial basic feasible solution. The process is cumbersome and may lead to more iterations in order to drive the artificial variables to zero. Sometimes it may be easier to start with an infeasible "better than optimal" basic solution and use the dual simplex method to achieve feasibility. It will be seen later that the dual simplex method is also helpful in certain aspects of sensitivity analysis.

In this method, we start with an infeasible but "better than optimal" basic solution; hence the coefficients of the nonbasic variables in equation (0) satisfy the optimality condition stated at the end of Section 6.1. This initial basis must be *dual-feasible*, that is, it must satisfy the feasibility condition for the dual of the given problem. The method, then, is one which maintains dual feasibility while removing the primal infeasibility.

We illustrate the method with an example which we have already considered.

Minimize z_y, with:

$$(0) \qquad z_y - 13y_1 - 17y_2 - 13y_3 \qquad\qquad = 0;$$

$$(1) \qquad\quad - 3y_1 \qquad - 2y_3 + \underline{y_4} \qquad\quad = -1;$$

$$(2) \qquad\quad - 2y_1 - y_2 - y_3 \qquad + \underline{y_5} \qquad = -3;$$

$$(3) \qquad\qquad\qquad - 3y_2 - y_3 \qquad\qquad + \underline{y_6} = -4.$$

Notice the signs in equations (1), (2), and (3).

Step 1. Select the basic variable y_k which is the most negative as the leaving basic variable. Here $k = 6$, *i.e.*, we select y_6.

Step 2. Take the ratios of the coefficients of the nonbasic variables of the current equation (0) to the corresponding negative coefficients in equation (k), the equation which contains the leaving basic variable. Do not consider zero or positive denominators. Here we take $-17/-3$ and $-13/-1$. The entering basic variable is the one whose coefficient yields the smallest ratio. This is the variable whose coefficient in equation (0) reaches 0 first as an increasing multiple of equation (k) is added to equation (0). Here y_2 is the entering basic variable, because $-17/-3 < -13/-1$.

Step 3. The pivoting process to change the basis is the same as the standard simplex method.

Step 4. If all the basic variables are now nonnegative, the solution must be feasible and optimal. Stop. If some are still negative, repeat the procedure.

To illustrate, the second set of equations is:

(0) $z_y - 13y_1 \quad -\frac{22}{3}y_3 \qquad\qquad -\frac{17}{3}y_6 = \frac{68}{3}$;

(1) $-3y_1 \quad -2y_3 + y_4 \qquad\qquad = -1$;

(2) $-2y_1 \quad -\frac{2}{3}y_3 \qquad + y_5 -\frac{1}{3}y_6 = \frac{-5}{3}$;

(3) $\qquad\qquad y_2 + \frac{1}{3}y_3 \qquad\qquad -\frac{1}{3}y_6 = \frac{4}{3}$.

y_5 is the next leaving variable. The ratios are $-13/-2$, $(-22/3)/(-2/3)$, and $(-17/3)/(-1/3)$. Hence y_1 is the entering variable. The next set of equations is:

(0) $z_y \qquad -3y_3 \qquad -\frac{13}{2}y_5 - \frac{7}{2}y_6 = \frac{67}{2}$;

(1) $-y_3 + y_4 - \frac{3}{2}y_5 + \frac{1}{2}y_6 = \frac{3}{2}$;

(2) $y_1 \quad +\frac{1}{3}y_3 \qquad -\frac{1}{2}y_5 + \frac{1}{6}y_6 = \frac{5}{6}$;

(3) $y_2 + \frac{1}{3}y_3 \qquad\qquad -\frac{1}{3}y_6 = \frac{4}{3}$.

The basic variables y_1, y_2 and y_4 are all nonnegative, so the iteration stops.

To use the dual simplex method on a maximization problem, the coefficients of the nonbasic variables in equation (0) should be all nonnegative. Step 1 remains unchanged, but Step 2 should be modified to select as entering variable the one whose coefficient gives the maximum ratio. Since the coefficients in row (0) are nonnegative while the denominators of the ratios are negative, this is the same as saying choose the ratio with the smallest absolute value.

6.5 Sensitivity analysis[1]

The constants c_j, b_i, and a_{ij}, better called *parameters*, of a linear programming problem are seldom determined with certainty. Even if they are, it may be necessary to change the values of some of them after an optimal solution has already been obtained. In order to avoid going through the simplex iterations from the very beginning, we perform a sensitivity analysis to determine the effect of some of the changes on the optimal solution. We shall not examine the situation where several changes take place simultaneously. Instead, we limit ourselves to single changes at a time. The following simple illustrative example will be used throughout this section unless otherwise stated. Let the primal problem be:
Maximize

$$z_x = 3x_1 + x_2,$$

---

[1]G. Thompson (1971) has some good, elementary discussions of the economic implications of changes in the parameters.

subject to

$$x_2 \leqslant 10,$$

$$2x_1 + x_2 \leqslant 12,$$

$$x_1 \geqslant 0, \qquad x_2 \geqslant 0.$$

The dual problem is:
Minimize

$$z_y = 10y_1 + 12y_2,$$

subject to

$$2y_2 \geqslant 3,$$

$$y_1 + \qquad y_2 \geqslant 1,$$

$$y_1 \geqslant 0, \qquad y_2 \geqslant 0.$$

After introducing slack variables, the initial and final sets of equations for the primal problem are:

(0)	$z_x - 3x_1 - x_2 \qquad\qquad = 0$	
(1)	$x_2 + x_3 \qquad = 10$	(6.5.1)
(2)	$2x_1 + x_2 \qquad + x_4 = 12$	

and

(0)	$z_x \qquad + \tfrac{1}{2}x_2 \qquad + \tfrac{3}{2}x_4 = 18$	
(1)	$x_2 + x_3 \qquad = 10$	(6.5.2)
(2)	$x_1 + \tfrac{1}{2}x_2 \qquad + \tfrac{1}{2}x_4 = 6$	

Therefore the optimal solution for the primal is $x_1^o = 6$, $x_2^o = 0$, $x_3^o = 10$, $x_4^o = 0$; the optimal solution for the dual is $y_1^o = 0$, $y_2^o = \tfrac{3}{2}$, $y_3^o = 0$, $y_4^o = \tfrac{1}{2}$; and $z_x^o = z_y^o = 18$.

6.5.1 Change in c_j when x_j^o is nonbasic

Suppose the coefficient of x_j in the initial equation (0) is changed from $-c_j$ to $-(c_j + \delta)$. In performing the simplex iterations, we add multiples of the constraint equations to (0), and only the coefficient of x_j in the final equation (0) is affected. Hence all we need to do is see if $(s_j^o - c_j - \delta)$ is still nonnegative. If it is, the solution is still optimal. If not, x_j will need to be introduced as an entering basic variable and the simplex method resumed from there.

To illustrate, suppose the coefficient of x_2 in (6.5.1) is changed from -1 to $-(1 + \delta)$. Then the coefficient of x_2 in (6.5.2) becomes $(\tfrac{1}{2} - \delta)$. As long as $\delta \leqslant \tfrac{1}{2}$, the original optimal solution still holds, otherwise x_2 should be introduced into the basis and the simplex iterations resumed until a new optimal solution is arrived at.

From the point of view of the dual problem, changing c_j to $c_j - \delta$ amounts to changing constraint j in the dual, and asking whether the primal feasible solution is still optimal is equivalent to asking whether the dual optimal solution is still feasible. In our example, the second constraint in the dual problem is changed to

$$y_1 + y_2 \geqslant 1 + \delta.$$

To find out whether the dual optimal solution is still feasible, we note that

$$0 + \tfrac{3}{2} \geqslant 1 + \delta$$

as long as $\delta \leqslant \tfrac{1}{2}$.

6.5.2 Change in c_j when x_j° is basic

If x_j is a basic variable, we want to make sure that its coefficient is 0 in the final equation (0). If $-c_j$ in the initial equation (0) is changed to $-(c_j + \delta)$, the final equation (0) will have $-\delta$ as the coefficient of x_j. To remove this coefficient, we multiply the (only)[2] equation that contains x_j in the final set of equations by δ and add the result to the final equation (0). Then check the resulting coefficients of the nonbasic variables in the new equation (0) to determine if any should become negative.

To illustrate, suppose equation (0) in (6.5.1) is

$$z_x - (3 + \delta)x_1 - x_2 = 0.$$

The new final equation (0) is

$$z_x + \left(\tfrac{1}{2} + \tfrac{1}{2}\delta\right)x_2 + \left(\tfrac{3}{2} + \tfrac{1}{2}\delta\right)x_4 = 18 + 6\delta.$$

Thus for $\delta \geqslant -1$, the original solution is still optimal although the optimum value of the objective function is changed by 6δ. If $\delta < -1$, we resume the simplex iterations to obtain a new optimal solution.

In the dual analysis, not only is the jth dual constraint changed, but the dual solution complementary to the previously optimal primal solution will also be changed since some of the s_{n+i}° are changed. In our example, the first dual constraint is changed to

$$2y_2 \geqslant 3 + \delta$$

and the new dual optimal solution is $y_1^\circ = 0$, $y_2^\circ = \tfrac{3}{2} + \tfrac{1}{2}\delta$, $y_3^\circ = 0$, $y_4^\circ = \tfrac{1}{2} + \tfrac{1}{2}\delta$. We need to check whether these new optimal values satisfy all the dual constraints. The second constraint, $y_1 + y_2 \geqslant 1$, is satisfied if $0 + \tfrac{3}{2} + \tfrac{1}{2}\delta \geqslant 1$, that is, if $\delta \geqslant -1$.

6.5.3 Change in b_i

Changing some b_i to $b_i + \delta$ will only change the right-hand sides of the equations and will not affect the coefficients of the variables in the final

---

[2]Why?

equation (0). Since the right-hand side is the value of the objective function or the value of a basic variable, we need to check whether the basic variables are still feasible, *i.e.*, whether they are nonnegative. If the new basis is feasible, the new solution is optimal.

Fortunately, x_{n+i}, the slack variable in equation (i), does not appear in any of the other initial equations and is equal to b_i. At any iteration the coefficients of δ must be the same row by row as the coefficients of x_{n+i}. This is also true for the final set of equations. So, if b_k^o is the right-hand side of the final equation (k), for $k=0,1,2,\ldots,m$, and $a_{k,n+i}$ is the coefficient of x_{n+i} in that equation, the new right-hand side after changing b_i to $(b_i+\delta)$ is $b_k^o+a_{k,n+i}^o\delta$.

To illustrate, if b_2 is changed to $12+\delta$, the new final set of equations becomes

(0) $$z_x \quad +\tfrac{1}{2}x_2 \quad +\tfrac{3}{2}x_4=18+\tfrac{3}{2}\delta,$$

(1) $$x_2+ x_3 \quad =10+0\cdot\delta,$$

(2) $$x_1+\tfrac{1}{2}x_2 \quad +\tfrac{1}{2}x_4= 6+\tfrac{1}{2}\delta.$$

For feasibility, all the right-hand sides must be nonnegative, or, $6+\tfrac{1}{2}\delta \geq 0$, or, $\delta \geq -12$. The new optimal solution needs to be revised so that $x_1^o=6+\tfrac{1}{2}\delta$, $x_2^o=0$, $x_3^o=10$, $x_4^o=0$, and the new $z_x^o=18+\tfrac{3}{2}\delta$.

For the dual problem we now test for optimality, since the new b_i's are the new coefficients in equation (0) of the dual. In our example, the new final equation (0) of the dual problem is

$$z_y-10y_1-\left(6+\tfrac{1}{2}\delta\right)y_3=18+\tfrac{3}{2}\delta,$$

and the original solution is optimal as long as $6+\tfrac{1}{2}\delta \geq 0$.

The b's play the same role in the dual problem as the c's in its dual (the primal). A graphic interpretation may be illuminating. Consider the example illustrated by Figure 6.5.1.

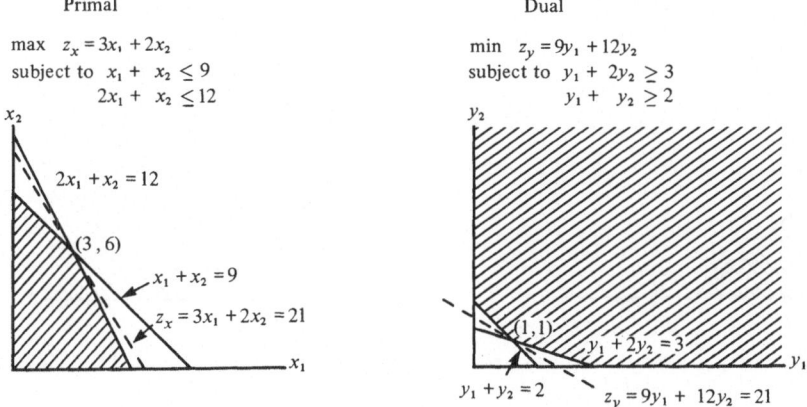

Primal

$$\max \ z_x = 3x_1 + 2x_2$$
subject to $x_1 + x_2 \leq 9$
$\quad\quad 2x_1 + x_2 \leq 12$

$2x_1 + x_2 = 12$

$(3,6)$

$x_1 + x_2 = 9$

$z_x = 3x_1 + 2x_2 = 21$

Dual

$$\min \ z_y = 9y_1 + 12y_2$$
subject to $y_1 + 2y_2 \geq 3$
$\quad\quad y_1 + y_2 \geq 2$

$(1,1)$

$y_1 + 2y_2 = 3$

$y_1 + y_2 = 2$ $z_y = 9y_1 + 12y_2 = 21$

Figure 6.5.1.

Suppose we change some b_i. Let us change b_1 from 9 to 6. The new problems are illustrated in Figure 6.5.2. Notice that when we change b_1, the slope of the line representing the primal objective function z_x remains the same but the feasible region is changed. In the dual problem, on the other hand, the feasible region remains the same but the slope of the line representing the dual objective function z_y is changed.

The slope of the line $z_y = b_1 y_1 + b_2 y_2$ depends, of course, on the vector

$$\mathbf{b} = \begin{pmatrix} b_1 \\ b_2 \end{pmatrix}.$$

In fact, the lines $z_y = b_1 y_1 + b_2 y_2$ are all perpendicular (orthogonal) to the vector $\mathbf{b}$ or some scalar multiple of $\mathbf{b}$. If the components of $\mathbf{b}$ are changed in such a way that the line $z_y = b_1 y_1 + b_2 y_2$ still lies in between the two boundary constraint lines represented by $a_{11} y_1 + a_{21} y_2 = c_1$ and $a_{12} y_1 + a_{22} y_2 = c_2$, and which intersect at the point (y_1^o, y_2^o), the original solution remains optimal.

The boundary constraint lines $a_{11} y_1 + a_{21} y_2 = c_1$ and $a_{12} y_1 + a_{22} y_2 = c_2$ are, again, orthogonal to the vectors

$$\begin{pmatrix} a_{11} \\ a_{21} \end{pmatrix} \quad \text{and} \quad \begin{pmatrix} a_{12} \\ a_{22} \end{pmatrix}$$

respectively. No loss of generality is incurred if we translate the constraint lines so that they meet at the origin, as shown in Figure 6.5.4, which is a translation of Figure 6.5.3. It is now apparent that the solution remains optimal if the vector $\mathbf{b}$ lies in between the two vectors

$$\begin{pmatrix} a_{11} \\ a_{21} \end{pmatrix} \quad \text{and} \quad \begin{pmatrix} a_{12} \\ a_{22} \end{pmatrix},$$

or in the cone spanned by them. The discussion can be extended to the

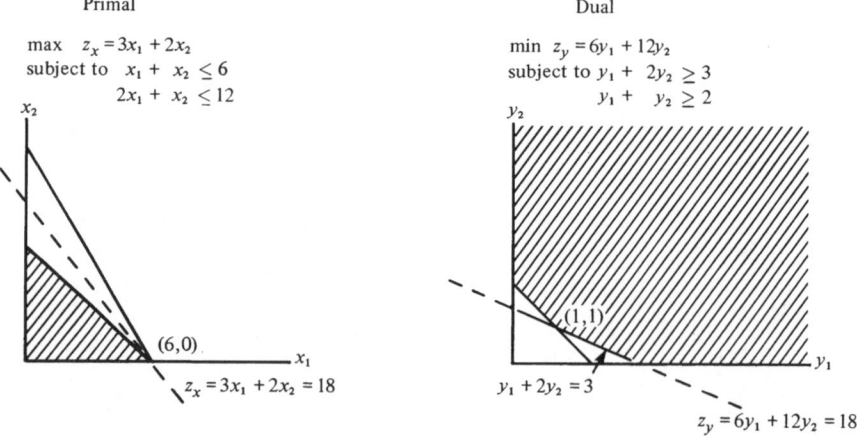

Figure 6.5.2.

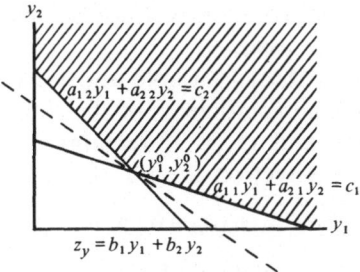

Figure 6.5.3.

general case where

$$\mathbf{b} \quad \text{and} \quad \begin{bmatrix} a_{1j} \\ \vdots \\ a_{mj} \end{bmatrix}$$

are $m \times 1$ vectors.

6.5.4 Change in a_{ij} when x_j^0 is nonbasic

Before we consider the effect of changing a_{ij}, let us see what happens if we introduce a new decision or activity variable, say x_5, to our example in (6.5.1) while keeping the parameters unchanged, and let it appear in equations (1) and (2) as (1), $+3x_5$ and (2), $+\frac{8}{3}x_5$. The question is: What coefficient c_5, when assigned to x_5 in the objective function, will necessitate entering x_5 into the basis?

Adding the column for x_5 creates the added constraint for the dual:

$$3y_1 + \tfrac{8}{3} y_2 \geqslant c_5.$$

Since the coefficients of the other variables, hence also the dual optimal solution, are unaffected, we may use the original dual solution. Substituting these current optimal dual variables in, we want

$$(3)(0) + \left(\tfrac{8}{3}\right)\left(\tfrac{3}{2}\right) \geqslant c_5.$$

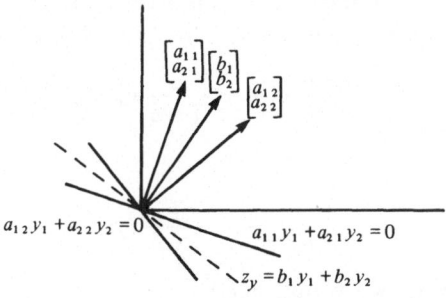

Figure 6.5.4.

Thus, if $c_5 \leqslant 4$, the optimal solution will not be changed. If $c_5 > 4$, then x_5 should be entered into the basis.

Similarly, if x_j^o is a nonbasic variable and we change a_{ij} to $a_{ij} + \delta$, the only possible change in the coefficients of the final equation (0) would be in the coefficient of x_j^o, and $s_{n+i}^o = y_i^o$ remains unchanged. Therefore we may simply examine the effect on the jth dual constraint, which becomes

$$a_{1j}y_1 + a_{2j}y_2 + \ldots + (a_{ij} + \delta)y_i + \ldots + a_{mj}y_m \geqslant c_j,$$

and substitute the optimal values of the y's into the above constraint to determine the range of δ for which x_j should or should not be entered as a basic variable in the primal.

To illustrate, suppose the coefficient of x_2 in equation (1) of (6.5.1) is changed to $1 + \delta$. The associated dual constraint with the optimal values of the dual variables is

$$(1 + \delta)(0) + (1)\left(\tfrac{3}{2}\right) \geqslant 1,$$

which is satisfied by all values of δ. Therefore changing that coefficient has no effect on the optimal solution. On the other hand, suppose, instead, that the coefficient of x_2 in equation (2) of (6.5.1) is the one that is changed to $1 + \delta$. The associated dual constraint with the optimal values of the dual variables is

$$(1)(0) + (1 + \delta)\left(\tfrac{3}{2}\right) \geqslant 1.$$

Now, if $\delta < \tfrac{-1}{3}$, it will be necessary to enter x_2 into the basis. If we look at the primal problem, we see that the final equation (0) is

$$z_x + \frac{1 + 3\delta}{2}x_2 + \tfrac{3}{2}x_4 = 18.$$

which also requires $1 + 3\delta \geqslant 0$, or, $\delta \geqslant \tfrac{-1}{3}$ for optimality.

6.5.5 Change in a_{ij} when x_j^o is basic

If x_j^o is basic, a change in its coefficient may change not only the previous solution but may affect the other coefficients in the final equation (0). We can no longer count on s_{n+i}^o to remain the same and therefore cannot use the original optimal dual solution to test for feasibility in the dual problem as we did when x_j^o was nonbasic. Let us illustrate the situation by changing the coefficient of x_1 in (6.5.1) to $2 + \delta$. The revised final set of equations is

$$(0) \qquad z_x + \qquad \frac{1 - \delta}{2 + \delta}x_2 + \qquad \frac{3}{2 + \delta}x_4 = \frac{36}{2 + \delta},$$

$$(1) \qquad\qquad\qquad\qquad x_2 + x_3 \qquad\qquad = 10,$$

$$(2) \qquad\qquad x_1 + \frac{1}{2 + \delta}x_2 + \qquad \frac{1}{2 + \delta}x_4 = \frac{12}{2 + \delta}.$$

To maintain optimality, we would need

$$-2 < \delta \leqslant 1.$$

Otherwise x_2 or x_4 will be entered into the basis. If $\delta \leqslant 2$, the original solution will not even be feasible. Notice also that the coefficient of x_4 in the final equation (0) is also changed, so that we can no longer say $y_2^o = \frac{3}{2}$. In fact, suppose we use the original optimal dual solution of $y_1^o = 0$, $y_2^o = \frac{3}{2}$, and use these values in the constraint

$$(2+\delta)y_2 \geqslant 3;$$

we find that $(2+\delta)(\frac{3}{2}) \geqslant 3$ holds for all $\delta \geqslant 0$, which is quite different from $-2 < \delta \leqslant 1$.

The situation involving a change in a_{ij} when x_j^o is basic is too complicated to be generalized in a few sentences.

For the dual problem, the jth constraint and the complementary dual solution will be affected. The new dual solution may or may not be optimal or feasible.

6.5.6 Addition of a new variable

Sometimes, after the optimal solution has been obtained, we may wish to add another activity variable. We have already commented on that situation in Subsection 6.5.4. The coefficient of the new variable in the initial equation (0) determines whether or not the associated new dual constraint will be satisfied, hence whether or not the original solution is optimal. Suppose it is not. Then it will be necessary to introduce that new variable into the basis.

Let us consider our example changed to

$$(0) \qquad\qquad z_x - 3x_1 - x_2 - 5x_3 \qquad\qquad = 0,$$

$$(1) \qquad\qquad\qquad\qquad x_2 + 3x_3 + \underline{x_4} \qquad = 10,$$

$$(2) \qquad\qquad 2x_1 + x_2 + \tfrac{8}{3}x_3 \qquad + \underline{x_5} = 12.$$

Notice that we have relabeled the slack variables because we now have $n = 3$. Substituting the original optimal solution of $y_1^o = 0$, $y_2^o = \frac{3}{2}$, into the new dual constraint

$$3y_1 + \tfrac{8}{3}y_2 \geqslant 5,$$

we see that

$$(3)(0) + \left(\tfrac{8}{3}\right)\left(\tfrac{3}{2}\right) \ngeqslant 5$$

and that this constraint is not satisfied—an outcome which, as we indicated earlier (Subsection 6.5.4), happens when $c_3 > 4$. It is necessary to enter x_3 into the basis. We take the original final set of equations in (6.5.2) and try to determine the new coefficients of x_3, keeping in mind that the variables labeled x_3 and x_4 in (6.5.1) and (6.5.2) are now labeled x_4 and x_5

respectively. Simple calculation yields

(0) $\qquad z_x \quad +\frac{1}{2}x_2+(-5+\frac{3}{2}\cdot\frac{8}{3})x_3 \qquad +\frac{3}{2}x_5=18,$

(1) $\qquad\qquad\qquad x_2+ \qquad\qquad 3x_3+x_4 \qquad\quad =10,$

(2) $\qquad \underline{x_1}+\frac{1}{2}x_2+ \qquad (\frac{1}{2})(\frac{8}{3})x_3 \qquad +\frac{1}{2}x_5= 6.$

We then enter x_3 into the basis and resume the simplex iterations from here.

In general, in order to find the coefficients of the added variable x_j in the revised final set of equations we must realize that the situation is as if that variable were originally in the initial set of equations, but with its initial coefficients c_j and a_{ij} all equal to 0. Therefore the same discussions concerning changes in c_j and a_{ij} of Subsections 6.5.1 and 6.5.4 apply. Let $a^o_{k,n+i}$ be the coefficient of x_{n+i} in the final (originally optimal) equation (k), for $k=1,2,\dots,m$. By the same reasoning that led to Theorem 6.1.1, we see that

$$a^o_{kj} = \sum_{i=1}^{m} a^o_{k,n+i}a_{ij} \quad \text{for } k=0,1,2,\dots,m.$$

Letting $s^o_j = a^o_{0j}$, the coefficient of x_j in the revised final equation (0) is again $s^o_j - c_j$, as in the previous sections.

6.5.7 Addition of a new constraint

The addition of a new constraint does not change any of the coefficients of the final equation (0), hence optimality is not affected. There is, however, the possibility that the feasibility region is reduced by the addition of a constraint and that the original solution is now "better than optimal" but infeasible. If that is the case, then we can start with the current solution and resort to the dual simplex method described in Section 6.4.

To illustrate, if we add a constraint

$$x_1+3x_2\leqslant 8$$

to our example, the original solution of $x^o_1=6$, $x^o_2=0$, remains feasible and optimal. However, if we add a constraint

$$x_1+x_2\geqslant 7,$$

the current optimal solution is no longer feasible. To solve the problem by the dual simplex method, we have

(0) $\qquad z_x \quad +\frac{1}{2}x_2 \qquad +\frac{3}{2}x_4 \qquad = 18,$

(1) $\qquad\qquad\qquad x_2+\underline{x_3} \qquad\qquad = 10,$

(2) $\qquad \underline{x_1}+\frac{1}{2}x_2 \qquad +\frac{1}{2}x_4 \qquad = 6,$

(3) $\qquad\qquad -x_1- x_2 \qquad\qquad + x_5 = -7.$

Since the basic variable x_1 should have a coefficient of 0 in equation (3), equation (2) is added to it to yield

(3) $$-\tfrac{1}{2}x_2 \qquad +\tfrac{1}{2}x_4 + x_5 = -1.$$

According to the dual simplex algorithm for maximization, x_2 is the entering basic variable, and the final set of equations is

(0) $$z_x \qquad\qquad +2x_4 + x_5 = 17,$$

(1) $$\underline{x_3} + x_4 + 2x_5 = 8,$$

(2) $$\underline{x_1} \qquad + x_4 + x_5 = 5,$$

(3) $$\underline{x_2} \qquad - x_4 - 2x_5 = 2.$$

The new optimal solution is $x_1^o = 5$, $x_2^o = 2$, $x_3^o = 8$, $x_4^o = 0$, $x_5^o = 0$, with $z_x^o = 17$.

Exercises

6.1
(a) Formulate the dual to the problem:
Minimize
$$z_y = y_1 + 3y_2,$$

subject to
$$4y_1 + y_2 \geqslant 8, \quad y_1 + y_2 \geqslant 5, \quad y_1 - 3y_2 \geqslant 1, \quad y_1, y_2 \geqslant 0.$$

(b) Find the optimal solutions to both problems.
(c) Write the combined tableau for both problems.

6.2 Change the third constraint in Problem 6-1 to $y_1 - 3y_2 \leqslant 1$, and repeat (a), (b), and (c) above.

6.3 Use the dual simplex method to solve Problem 6.1.

6.4 Use the dual simplex method to solve Problem 6.2.

6.5 Refer to (5.2.1). From Figure 5.2.1, we know that the optimal value of $z = x_1 + 4x_2$ is 22, with $x_1 = 2$ and $x_2 = 5$. Find the optimal solution if:

(a) The objective function is changed to $z = 3x_1 + 4x_2$;
(b) The second constraint is changed to $-3x_1 + 2x_2 \leqslant 25/3$.

6.6 Refer to (5.2.9), Table 5.3.1, and the dual simplex tableaux of Section 6.4. Determine the range within which each of the following parameters may be changed without our having to resume the simplex iterations:

(a) c_1, the coefficient of x_1;
(b) c_2, the coefficient of x_2;
(c) b_2 (which was 17 in (5.2.1));
(d) a_{11} (which was 3 in (5.2.1)).

6.7 A firm uses two inputs to produce three products. Let z_x = total profit from the three products and z_y = total value to be inputed to the resources.
(a) Formulate the primal and dual linear programs.
(b) Discuss the economic meaning of increasing by one unit one of the following while holding the others unchanged: (i) c_1; (ii) b_1; (iii) a_{12}.

7 Nonlinear Programming

7.1 General nonlinear programming problems

As commonly used, the terminology "nonlinear programming" refers to the problem of finding an optimal (maximal or minimal) value of a given function $f(x_1, x_2, \ldots, x_n)$ of n nonnegative variables $x_1, x_2, \ldots, x_n$ ($x_j \geq 0$ for $j = 1, 2, \ldots, n$) subject to m constraints of the type $g^i(x_1, x_2, \ldots, x_n)$ $\{\leq, =, \geq\}$ b_i, $i = 1, 2, \ldots, m$, where the b_i are constants and the functions f and g^i are not all linear, as they were in Chapters 5 and 6. It is apparent that the problem now becomes vastly more difficult than the linear programming one. The relaxation of the assumption of linearity greatly increases the difficulty in computation, and the scope and range of topics are much wider and more varied.

In linear programming, the following hold:

(1) The set of feasible solutions is a convex set with a finite number of corner (extreme) points;
(2) Since the objective function is linear, the different values of the function

$$f(x_1, \ldots, x_n) = \sum_{j=1}^{n} c_j x_j = z_k$$

for different k are hyperplanes parallel to each other;
(3) If the objective function is bounded, its optimal value will occur at one of the extreme points of the convex set of feasible solutions, possibly at more than one extreme point;
(4) A local optimum is also the global optimum of the objective function over the set of feasible solutions.

The simplex method is based on the above properties. In nonlinear programming, some or all of these properties may not hold. As an illustration, consider the problem:

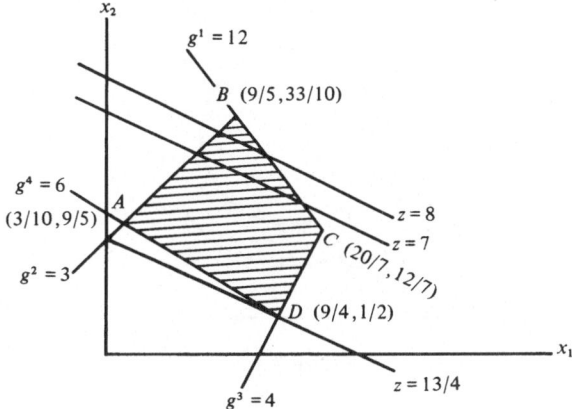

Figure 7.1.1.

Minimize

$$z = x_1 + 2x_2$$

subject to

(1)
$$g^1 = \quad 3x_1 + 2x_2 \leqslant 12,$$

(2)
$$g^2 = -2x_1 + 2x_2 \leqslant 3,$$

(3)
$$g^3 = \quad 2x_1 - \ x_2 \leqslant 4,$$

(4)
$$g^4 = \quad 2x_1 + 3x_2 \geqslant 6,$$

$$x_1 \geqslant 0, \qquad x_2 \geqslant 0.$$

(7.1.1)

This is a simple linear programming problem and the minimal value of z is $13/4$, as shown in Figure 7.1.1.

Suppose now we keep the same constraints (hence the same feasible region) but change the objective function to a nonlinear one, say

$$z = (x_1 - 4)^2 + (x_2 - 3)^2.$$

The minimal value of z no longer occurs at an extreme point of the feasible region but at a point on the boundary, with $x_1 = 34/13$, $x_2 = 27/13$, and $z = 468/169$, as shown in Figure 7.1.2.[1] Clearly the simplex method of examining the corner points will not help us here.

If the objective function is

$$z = \left(x_1 - \tfrac{9}{5}\right)^2 + (x_2 - 2)^2$$

---

[1] For the interested reader, we solved for the values of x_1 and x_2 and z by utilizing the fact that at the point of tangency, the slope of the circle $z = (x_1 - 4)^2 + (x_2 - 3)^2$ must be the same as the slope of the line $3x_1 + 2x_2 = 12$, which is $-3/2$. By implicit differentiation, we have $0 = 2(x_1 - 4) + 2(x_2 - 3)dx_2/dx_1$. Since $dx_2/dx_1 = -3/2$ is the slope of the circle at the point of tangency, substitute this into the equation to get $0 = 2(x_1 - 4) + 2(x_2 - 3)(-3/2)$, that is, $2x_1 - 3x_2 = -1$. Solving this equation simultaneously with $3x_1 + 2x_2 = 12$ yields the results.

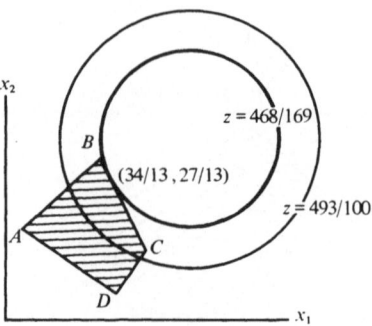

Figure 7.1.2.

and the constraints are the same, the minimal value of z occurs at the point $x_1 = 9/5$, $x_2 = 2$, which is an interior point of the feasible region. Obviously, $z = 0$ is the minimum. This is shown in Figure 7.1.3.

Figure 7.1.3 also shows that a local optimum is not always a global optimum. The function

$$z = \left(x_1 - \tfrac{9}{5}\right)^2 + (x_2 - 2)^2$$

attains a local maximum at the point B, but this is not the global maximum, which occurs at the point D.

The examples given so far all have the same linear constraints, thus the same convex set of feasible solutions. If some or all of the constraints are not linear, the set of feasible solutions may not even be convex, as shown by the disjoint shaded area in Figure 7.1.4, for which the constraints are:

$$x_1 x_2 \leqslant 1;$$

$$x_1^2 + x_2^2 \geqslant 4;$$

$$x_1 \geqslant 0,\ x_2 \geqslant 0.$$

Convexity, along with some other assumption, provides us with a theory which states that a local optimum is also a global optimum. Hence,

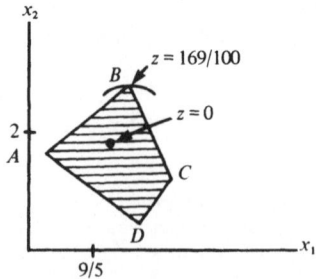

Figure 7.1.3.

Figure 7.1.4.

without convexity the difficulties encountered in solving such a programming problem are generally much greater.

There is no one algorithm, such as the simplex algorithm for linear
programming problems, which is applicable to a wide class of nonlinear
programming problems. Various computational methods and techniques
have been devised to solve the different problems, and a good summary of
many of them may be found in Aoki (1971), Converse (1970), and Hayes
(1975). No attempt will be made in this chapter to examine these numerous
techniques. Instead, we shall briefly discuss a few of the analytical ideas
that characterize an optimal solution, in particular the Kuhn–Tucker
theorem, which states a necessary condition for an optimum. The Kuhn–
Tucker conditions show us how to recognize whether a solution is an
optimal one or not, although they do not provide us with a computational
technique for obtaining such a solution, except in perhaps the simplest
cases. The theory that underlies these conditions is, however, a fundamental one which pervades all the approaches to the topic of optimization.

7.2 The Kuhn–Tucker conditions

Consider the general nonlinear programming problem:
Maximize
$$z = f(x_1, x_2, \ldots, x_n)$$
subject to

$$g^i(x_1, x_2, \ldots, x_n) \leq b_i \quad \text{for } i = 1, \ldots, u,$$

$$g^i(x_1, \ldots, x_n) \geq b_i \quad \text{for } i = u+1, \ldots, v, \tag{7.2.1}$$

$$g^i(x_1, \ldots, x_n) = b_i \quad \text{for } i = v+1, \ldots, m,$$

$$x_j \geq 0 \quad \text{for } j = 1, 2, \ldots, n,$$

where the functions f and g^i are not necessarily linear but have continuous
first derivatives and satisfy certain regularity conditions.[2] We shall again

---

[2]These are called the *constraint qualification* by Kuhn and Tucker in their original paper. (See
Kuhn and Tucker (1951)). In essence, the constraint qualification rules out rare situations on
the boundaries of the feasibility region which might invalidate the conditions stated in their
theorem. See A. Chiang (1974), pp. 713–20, for a good and elementary discussion of the
constraint qualification. See also Aoki (1971), pp. 179–187.

use the superscript "o" to indicate the values at the optimum point, and state the celebrated Kuhn–Tucker theorem as follows:

Theorem 7.2.1. *In order for* $\mathbf{x}^{\circ} = (x_1^{\circ}, x_2^{\circ}, \ldots, x_n^{\circ})^{T}$ *to be an optimal solution, it is necessary that, in addition to the feasibility conditions of* (7.2.1), *there exist* m *numbers* $\lambda_1, \lambda_2, \ldots, \lambda_m$ *such that the following conditions hold*:

(1) $(\partial f(\mathbf{x}^{\circ})/\partial x_j) - \sum_{i=1}^{m} \lambda_i^{\circ}(\partial g^{i}(\mathbf{x}^{\circ})/\partial x_j) = 0$ *for those components* x_j° *of* $\mathbf{x}^{\circ}$ *which are strictly positive, and* $(\partial f(\mathbf{x}^{\circ})/\partial x_j) - \sum_{i=1}^{m} \lambda_i^{\circ}(\partial g^{i}(\mathbf{x}^{\circ})/\partial x_j) \leq 0$ *for those components* $x_j^{\circ} = 0.$[3]   (7.2.2)

(2) $\lambda_i^{\circ} \geq 0$ *for* $i = 1, \ldots, u$, $\lambda_i^{\circ} \leq 0$ *for* $i = u+1, \ldots, v$, λ_i° *is unrestricted in sign for* $i = v+1, \ldots, m$, *and* $\lambda_i^{\circ} = 0$ *if constraint* i *is not active at the optimum point* $\mathbf{x} = \mathbf{x}^{\circ}$ (*A constraint* g^{k} *is active at the point* $\mathbf{x}^{\circ}$ *if* $g^{k}(\mathbf{x}^{\circ}) = 0$. *If strict inequality holds at* $\mathbf{x}^{\circ}$, *then that constraint is inactive and may be ignored at that point*).   (7.2.3)

(3) *If* $\lambda_i^{\circ} \neq 0$, *then* $b_i - g^{i}(\mathbf{x}^{\circ}) = 0$. *If* $\lambda_i^{\circ} = 0$, *then* $b_i - g^{i}(\mathbf{x}^{\circ}) \geq 0$ *for* $i = 1, \ldots, u$, *and* $b_i - g^{i}(\mathbf{x}^{\circ}) \leq 0$ *for* $i = u+1, \ldots, v$.   (7.2.4)

The three conditions above imply

(1′)
$$x_j^{\circ}\left[\frac{\partial f(\mathbf{x}^{\circ})}{\partial x_j} - \sum_{i=1}^{m} \lambda_i^{\circ}\frac{\partial g^{i}(\mathbf{x}^{\circ})}{\partial x_j}\right] = 0 \quad \text{for } j = 1, \ldots, n,$$
(7.2.5)

and

(3′) $\lambda_i^{\circ}[b_i - g^{i}(\mathbf{x}^{\circ})] = 0 \quad \text{for } i = 1, \ldots, m.$

For minimization, change the second half of the condition given by (7.2.2) to read "$\partial f(\mathbf{x}^{\circ})/\partial x_j) - \sum_{j=1}^{m} \lambda_i^{\circ}(\partial g^{i}(\mathbf{x}^{\circ})/\partial x_j) \geq 0$ for those $x_j^{\circ} = 0$," and reverse the sense of the inequalities in (7.2.3). We use the notation λ_i° to indicate those values of λ_i which correspond to the optimal point $\mathbf{x}^{\circ}$.

For brevity, we shall refer to the Kuhn–Tucker conditions as K–T conditions.

Example 1. As an illustration of the K–T conditions, let us consider the problem cited in Section 7.1, namely:
Maximize

$$z = f(x_1, x_2) = \left(x_1 - \tfrac{9}{5}\right)^{2} + (x_2 - 2)^{2}$$

with the constraints stated in (7.1.1), that is,

$$g^{1}(x_1, x_2) = 3x_1 + 2x_2 \leq 12,$$

$$g^{2}(x_1, x_2) = -2x_1 + 2x_2 \leq 3,$$

$$g^{3}(x_1, x_2) = 2x_1 - x_2 \leq 4,$$

$$g^{4}(x_1, x_2) = 2x_1 + 3x_2 \geq 6,$$

$$x_1 \geq 0, \qquad x_2 \geq 0.$$

---

[3]By $\partial f(\mathbf{x}^{\circ})/\partial x_j$ and $\partial g^{i}(\mathbf{x}^{\circ})/\partial x_j$ we mean the partial derivative at $x_j = x_j^{\circ}, j = 1, 2, \ldots, n.$

From Section 7.1, we know that the optimum point is at $D(9/4, 1/2)$, which is the intersection of g^3 and g^4, with $z = 981/400$. At the optimum point, constraints g^1 and g^2 are inactive. Hence, according to the K–T condition given by (7.2.3), we expect $\lambda_1 = \lambda_2 = 0$, $\lambda_3 \geqslant 0$, and $\lambda_4 \leqslant 0$.

$$\frac{\partial f(\mathbf{x}^o)}{\partial x_1} = 2\left(x_1 - \tfrac{9}{5}\right)\big|_{x_1 = 9/4} = \tfrac{9}{10},$$

$$\frac{\partial f(\mathbf{x}^o)}{\partial x_2} = 2(x_2 - 2)\big|_{x_2 = 1/2} = -3.$$

Since $x_1^o > 0$ and $x_2^o > 0$, (7.2.2) states that $(\partial f(\mathbf{x}^o)/\partial x_j) - \sum_{i=1}^4 \lambda_i^o (\partial g^i(\mathbf{x}^o)/\partial x_j)$ $= 0$, for $j = 1, 2$. Therefore,

$$\tfrac{9}{10} = 0 + 0 + 2\lambda_3 + 2\lambda_4,$$

$$-3 = 0 + 0 - \lambda_3 + 3\lambda_4.$$

Solving, we get $\lambda_3 = 87/80$, $\lambda_4 = -51/80$.

Condition (7.2.4) is easily shown to be satisfied, so we have found the four numbers λ_i which satisfy the K–T conditions.   □

It must be emphasized that the K–T conditions are necessary, but not sufficient, for the optimum solution (global maximum or minimum)[4] to the general nonlinear programming problem. In order to get sufficiency, the feasible region of solutions needs to be convex and the objective function must be concave (for maximum) or convex (for minimum).[5] Let us state this as a corollary to Theorem 7.2.1:

Corollary 7.2.1. *If the functions $f(\mathbf{x})$ and $g^i(\mathbf{x})$ of Theorem 7.2.1 are such that $f(\mathbf{x})$ is a concave (convex) function, and $g^i(\mathbf{x})$ is convex if $\lambda_i^o > 0$ and concave if $\lambda_i < 0$, then the K–T conditions are necessary and sufficient for $f(\mathbf{x}^o)$ to be the global maximum (minimum).*   □

Example 2. Consider the same problem of Example 1, in which the optimum point is at $(9/4, 1/2)$. Since the constraints are linear, they are both concave and convex. Suppose we try the point $B(9/5, 33/10)$, although we know from the last section that it does not yield the global maximum of the function. ($z = 981/400$ at the point D, but $z = 169/100$ at B.)

$$\left.\left|\frac{\partial f}{\partial x_1}\right|\right._{x_1 = 9/5} = 0, \qquad \left.\left|\frac{\lambda f}{\lambda x_2}\right|\right._{x_2 = 33/10} = \tfrac{13}{5}.$$

At the point $(9/5, 33/10)$, constraints g^3 and g^4 are inactive, so we would expect $\lambda_1 \geqslant 0$, $\lambda_2 \geqslant 0$, $\lambda_3 = \lambda_4 = 0$.

[4]Actually, the K–T conditions apply to a local extremum, but a global extremum is also a local extremum.

[5]See Appendix II for a definition of convex or concave functions. With some other qualifications, the convexity or concavity restrictions may be relaxed to quasiconvex or quasiconcave restrictions. See Zangwill (1969).

In solving the two equations $3\lambda_1 - 2\lambda_2 = 0$ and $2\lambda_1 + 2\lambda_2 = 13/5$, we get $\lambda_1 = 13/25$ and $\lambda_2 = 39/50$.

So we have found the λ's which satisfy the Kuhn–Tucker conditions. However, the function z is not concave, although the functions g^i are convex. Therefore the conditions are not sufficient to make $(9/5, 33/10)$ the optimum point.   □

Example 3. Let us now use the same constraints as in the previous two examples but let the objective function be

$$z = 3x_1 + x_2.$$

This is a simple linear programming problem (a special case of the general problem) and it is easy to see that the maximum value of z is attained at the point $C(20/7, 12/7)$, which is the intersection of g^1 and g^3.

We apply the K–T conditions to obtain the λ's as follows:

$$\frac{\partial z}{\partial x_1} = 3, \qquad \frac{\partial z}{\partial x_2} = 1, \quad \text{and} \quad \lambda_2 = \lambda_4 = 0;$$

$$3\lambda_1 + 2\lambda_3 = 3,$$

$$2\lambda_1 - \lambda_3 = 1,$$

and $\lambda_1 = 5/7$, $\lambda_3 = 3/7$, both of which are nonnegative, as expected. Since the function z, being linear, is convex and concave and we have found the λ's which satisfy all of the K–T conditions, we know that the point $(20/7, 12/7)$ is indeed the optimum point.

Suppose we try the point $(9/5, 33/10)$ and ask if it could be the optimum point. If it were we ought to be able to find the four λ's which satisfy the K–T conditions, with $\lambda_1 \geqslant 0$, $\lambda_2 \geqslant 0$, $\lambda_3 = \lambda_4 = 0$. In solving the two equations

$$3\lambda_1 - 2\lambda_2 = 3 \qquad 2\lambda_1 + 2\lambda_2 = 1,$$

we get $\lambda_1 = 4/5$, but $\lambda_2 = -3/10$, a negative number. So, happily for us, the K–T conditions are not satisfied. If they were satisfied, that would make $(9/5, 33/10)$ the optimum point, because in this case the conditions are both necessary and sufficient.   □

7.3 Further look at the Kuhn–Tucker conditions

We shall not look at the exact formulation of the K–T conditions or at their proof as given by the authors. The interested reader is referred to Kuhn–Tucker (1951) and to Hadley (1964), Chapter 6, for a good explanation. Instead, we shall show the connection between the K–T conditions and the classical approach to maximization (minimization) on the one hand and duality in linear programming on the other.

7.3.1 Nonnegativity and inequality constraints

It will be recalled that in Chapter 3, which deals with (classical) maximization with constraints, we formed the Lagrangian function

$$L(\mathbf{x}, \boldsymbol{\lambda}) = f(\mathbf{x}) + \sum_{i=1}^{m} \lambda_i [b_i - g^i(\mathbf{x})],$$

where $\mathbf{x} = (x_1, x_2, \ldots, x_n)^T$ and $\boldsymbol{\lambda} = (\lambda_1, \lambda_2, \ldots, \lambda_m)^T$, and stated that a necessary condition for $f(\mathbf{x}^\circ) = f(x_1^\circ, x_2^\circ, \ldots, x_n^\circ)$ to be a maximum (minimum) is that there exist m numbers λ_i, $i = 1, \ldots, m$, such that

$$\frac{\partial L(\mathbf{x}^\circ, \boldsymbol{\lambda})}{\partial x_j} = \frac{\partial f(\mathbf{x}^\circ)}{\partial x_j} - \sum_{i=1}^{m} \lambda_i \frac{\partial g^i(\mathbf{x}^\circ)}{\partial x_j} = 0$$

and

$$\frac{\partial L(\mathbf{x}^\circ, \boldsymbol{\lambda})}{\partial \lambda_i} = b_i - g^i(\mathbf{x}^\circ) = 0.$$

There were no restrictions on the sign of the x_j's or the λ_i's, and the constraints were in the form of equations. Suppose we now add the nonnegativity requirements of $x_j \geq 0$, $j = 1, \ldots, n$, and allow inequality constraints. The situation then becomes one which we have already considered, that is to say, a situation involving nonlinear programming. Again we assume that the objective function f and the constraint functions g^i have continuous first derivatives and satisfy the regularity conditions stated before Theorem 7.2.1.

The inequality constraints can be made to become equations by the introduction of slack or surplus variables x_{si}, with the requirement that $x_{si} \geq 0$. Thus, if the problem has m constraints, with

$$g^i(\mathbf{x}) \leqslant b_i \quad \text{for } i = 1, \ldots, u,$$

$$g^i(\mathbf{x}) \geqslant b_i \quad \text{for } i = u+1, \ldots, v,$$

and

$$g^i(\mathbf{x}) = b_i \quad \text{for } i = v+1, \ldots, m,$$

the constraints are equivalent to

$$g^i(\mathbf{x}) + x_{si} = b_i \quad \text{for } i = 1, \ldots, u,$$

$$g^i(\mathbf{x}) - x_{si} = b_i \quad \text{for } i = u+1, \ldots, v,$$

and

$$g^i(\mathbf{x}) = b_i \quad \text{for } i = v+1, \ldots, m,$$

$$x_{si} \geqslant 0 \quad \text{for } i = 1, \ldots, v.$$

Let us form the expanded Lagrangian function

$$L(\mathbf{x},\mathbf{x}_s,\boldsymbol{\lambda})=f(\mathbf{x})+\sum_{i=1}^{u}\lambda_i[b_i-x_{si}-g^i(\mathbf{x})]+\sum_{i=u+1}^{v}\lambda_i[b_i+x_{si}-g^i(\mathbf{x})]$$

$$+\sum_{i=v+1}^{m}\lambda_i[b_i-g^i(\mathbf{x})],$$

where $\mathbf{x}_s=(x_{s1},x_{s2},\ldots,x_{sv})^T$. It is easy to see that the λ's are generalized Lagrange multipliers of classical optimization.

Next, we consider the nonnegativity requirement. Recall from Chapter 1 that when we derived the necessary condition for a maximum (minimum), which states that the first derivative (slope) should be equal to zero, we were restricting ourselves to interior points. As illustration, take a function of only one variable, say $f(x)$. Suppose that the maximum point in question is a boundary point with $x=0$, such as points A or B in Figure 7.3.1(a). While the slope of the function at B is 0, the slope at A is not 0 but is negative. Similarly, as seen in Figure 7.3.1(b), the slope of a boundary minimum point may be either 0 or positive. On the other hand, if the maximum (minimum) occurs at an interior point, then the first derivative must be 0.

Returning to our general nonlinear programming problem and concentrating on the nonnegative *orthant* (in which all the component x's, including the slack and surplus variables, are nonnegative), we shall consider these two situations: (1) The maximum (minimum) occurs at an interior point; (2) One or more of the components x_j^o of the optimum point $\mathbf{x}^o=(x_1^o,x_2^o,\ldots,x_n^o)^T$ is (are) equal to zero.

(1) At an interior point where $x_j^o>0$ and $x_{si}^o>0$ we take the expanded Lagrangian and solve for all the equations

$$\frac{\partial L}{\partial x_j}(\mathbf{x}^o,\mathbf{x}_s^o,\boldsymbol{\lambda}^o)=\frac{\partial f(\mathbf{x}^o)}{\partial x_j}-\sum_{i=1}^{m}\lambda_i\frac{\partial g^i(\mathbf{x}^o)}{\partial x_j}=0,\qquad j=1,\ldots,n,$$

$$\frac{\partial L}{\partial \lambda_i}(\mathbf{x}^o,\mathbf{x}_s^o,\boldsymbol{\lambda}^o)=b_i-g^i(\mathbf{x}^o)\pm x_{si}^o=0,\qquad i=1,\ldots,m,\qquad(7.3.1)$$

$$x_{si}^o=0,\qquad i=v+1,\ldots,m,$$

$$\frac{\partial L}{\partial x_{si}}(\mathbf{x}^o,\mathbf{x}_s^o,\boldsymbol{\lambda}^o)=\pm\lambda_i^o=0,\qquad i=1,\ldots,v,$$

where $\boldsymbol{\lambda}^o=(\lambda_1^o,\ldots,\lambda_m^o)^T$ and $\mathbf{x}_s^o=(x_{s1}^o,\ldots,x_{sv}^o,0,\ldots,0)^T$ are the vectors $\boldsymbol{\lambda}$ and $\mathbf{x}_s$ corresponding to the optimum $\mathbf{x}^o$.

The first line of (7.3.1) is simply the first line of (7.2.2); and the last line of (7.3.1) repeats the statement of (7.2.3) that $\lambda_i^o=0$ if constraint i is inactive at $\mathbf{x}^o$, because $x_{si}^o>0$ means precisely that.

Moreover, since $x_{si}^o=\pm(b_i-g^i(\mathbf{x}^o))$, depending on whether the ith constraint has $g^i\leqslant$ or $\geqslant b_i$,

$$\lambda_i^o\cdot x_{si}^o=0$$

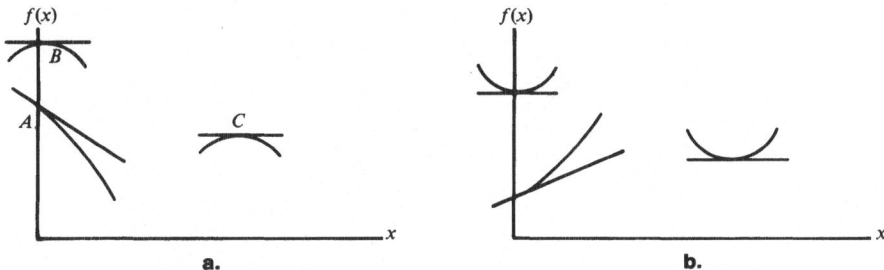

Figure 7.3.1

is the same as

$$\lambda_i^\circ[b_i - g^i(x^\circ)] = 0,$$

which is (3') of the K–T conditions.

(2) At a boundary point, as an example, suppose we have a function $z = f(x_1, x_2)$ under the constraints $g(x_1, x_2) = b$ and $x_1 \geqslant 0$, $x_2 \geqslant 0$. The graph is shown in Figure 7.3.2. We have a family of curves $z = f(x_1, x_2)$ intersecting the curve $g(x_1, x_2) = b$. The function attains its maximum at P_2 (where the second coordinate $x_2 = 0$) and minimum at P_1 (where the first coordinate $x_1 = 0$). If we form the Lagrangian

$$L(x_1, x_2, \lambda) = f(x_1, x_2) + \lambda[b - g(x_1, x_2)],$$

and set $x_1 = x_1^{\max}$ and $\lambda = \lambda^{\max}$ (that value of λ which corresponds to the point P_2), L becomes a function of x_2 alone. The graph of L near P_2, where $x_2 = x_2^{\max} = 0$, will appear as in Figure 7.3.3. At P_2, the slope of L (its derivative with respect to x_2 at $x_2^{\max} = 0$) is not necessarily 0 but is $\leqslant 0$. On the other hand, if we set $x_1 = x_1^{\min}$ and $\lambda = \lambda^{\min}$, L is again a function of x_2 alone. But in a neighborhood of P_1, where $x_2^{\min} > 0$, the graph of L will look quite different from that in Figure 7.3.3. It will look something like that shown in Figure 7.3.4, and the slope of L will be equal to 0 at $x_2 = x_2^{\min}$. Thus the slope of $L \leqslant 0$ in either case. Similar graphs may be drawn for L as a function of x_1 alone by setting x_2 and λ_2 equal to the corresponding constants $x_2^{\min}$ (or $x_2^{\max}$) and $\lambda_2^{\min}$ (or $\lambda_2^{\max}$).

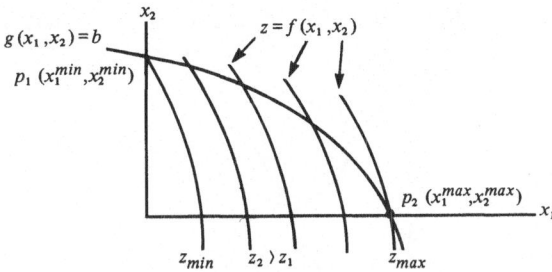

Figure 7.3.2.

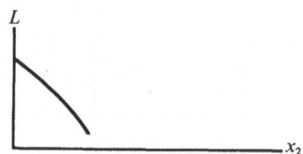

Figure 7.3.3.

The same analysis can be applied to a function of more than two variables, from which it will be seen that in order for a function to attain its maximum in the nonnegative orthant, it is necessary, if $x_j^o = 0$, that for some $j, j = 1, \ldots, n$, we must have

$$\frac{\partial L}{\partial x_j}(\mathbf{x}^o, \mathbf{x}_s, \boldsymbol{\lambda}^o) = \frac{\partial f(\mathbf{x}^o)}{\partial x_j} - \sum_{i=1}^{m} \lambda_i \frac{\lambda g^i(\mathbf{x}^o)}{\partial x_j} \leqslant 0, \qquad (7.3.2)$$

which is the second line of (7.2.2).

7.3.2 Saddle point

Before we explore the close relationship between the K–T conditions and duality in linear programming, we shall engage in a small digression on the concept of a saddle point.

A function $F(\mathbf{x}, \boldsymbol{\lambda})$, where $\mathbf{x} = (x_1, \ldots, x_n)^T$ and $\boldsymbol{\lambda} = (\lambda_1, \ldots, \lambda_m)^T$, is said to have a *saddle point* at $\mathbf{x}^o = (x_1^o, \ldots, x_n^o)^T$ and $\boldsymbol{\lambda}^o = (\lambda_1^o, \ldots, \lambda_m^o)^T$ if there exists an ε-neighborhood of $(\mathbf{x}^o, \boldsymbol{\lambda}^o)$ such that for all $(\mathbf{x}, \boldsymbol{\lambda})$ in that neighborhood we have

$$F(\mathbf{x}, \boldsymbol{\lambda}^o) \leqslant F(\mathbf{x}^o, \boldsymbol{\lambda}^o) \leqslant F(\mathbf{x}^o, \boldsymbol{\lambda}).$$

That is to say, F, as a function of the vectors $\mathbf{x}$ and $\boldsymbol{\lambda}$, has a maximum with respect to $\mathbf{x}$ at the point $(\mathbf{x}^o, \boldsymbol{\lambda}^o)$ but a minimum with respect to $\boldsymbol{\lambda}$.

It can be proved that the K–T conditions are necessary for our Lagrangian function

$$L(\mathbf{x}, \boldsymbol{\lambda}) = f(\mathbf{x}) + \sum_{i=1}^{m} \lambda_i [b_i - g^i(\mathbf{x})]$$

to have a *degenerate* form of saddle point at $(\mathbf{x}^o, \boldsymbol{\lambda}^o)$, that is, to obey the

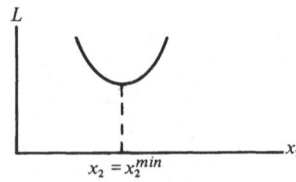

Figure 7.3.4.

inequality

$$L(\mathbf{x}, \boldsymbol{\lambda}^\circ) \leqslant L(\mathbf{x}^\circ, \boldsymbol{\lambda}^\circ) = L(\mathbf{x}, \boldsymbol{\lambda}^\circ).^6 \qquad (7.3.3)$$

Our previous discussions do not constitute a proof, but make it easy to see that the K–T conditions are necessary for the left-hand inequality, that is, for $L(\mathbf{x}^\circ, \boldsymbol{\lambda}^\circ)$ to be the maximum with respect to $\mathbf{x}$, with $\boldsymbol{\lambda}$ fixed at $\boldsymbol{\lambda}^\circ$.

To see that it is a minimum with respect to $\boldsymbol{\lambda}$, we may reason as follows.

We assume that $f(\mathbf{x})$ has a maximum with respect to $\mathbf{x}$ for those $\mathbf{x}$'s satisfying the constraints $g^i(\mathbf{x}) = b_i$ $(i = 1, \ldots, m)$ at the point $\mathbf{x}^\circ$ and that $\boldsymbol{\lambda}^\circ = (\lambda_1^\circ, \lambda_2^\circ, \ldots, \lambda_m^\circ)^{\mathrm{T}}$ is the vector of Lagrange multipliers which corresponds to this optimizing point $\mathbf{x}^\circ$. We now consider the Lagrangian function $L(\mathbf{x}, \boldsymbol{\lambda}) = f(\mathbf{x}) + \sum_{i=1}^m \lambda_i [b_i - g^i(\mathbf{x})]$ in general. $L(\mathbf{x}, \boldsymbol{\lambda})$ coincides with $f(\mathbf{x})$ if $\mathbf{x}$ satisfies the constraints $g^i(\mathbf{x}) = b_i$; otherwise L and f are not the same. It is often possible that there exists an ε-neighborhood of $\mathbf{x}^\circ$ such that $L(\mathbf{x}, \boldsymbol{\lambda}^\circ)$, as a function of $\mathbf{x}$ alone (since $\boldsymbol{\lambda}$ is fixed at $\boldsymbol{\lambda}^\circ$), has an unconstrained local maximum at $\mathbf{x}^\circ$ with respect to all the $\mathbf{x}$'s in that ε-neighborhood, whether or not they satisfy the constraints $g^i(\mathbf{x}) = b_i$ $(i = 1, \ldots, m)$. That is,

$$L(\mathbf{x}, \boldsymbol{\lambda}^\circ) \leqslant L(\mathbf{x}^\circ, \boldsymbol{\lambda}^\circ), \qquad (7.3.4)$$

where the $\mathbf{x}$'s are unconstrained.

Next, assume that there exists a δ-neighborhood of $\boldsymbol{\lambda}^\circ$ (in m-space) such that, for each $\boldsymbol{\lambda}$ in that δ-neighborhood, the function $L(\mathbf{x}, \boldsymbol{\lambda})$ has an unconstrained maximum with respect to $\mathbf{x}$ at a point $(\hat{\mathbf{x}}, \boldsymbol{\lambda})$ which satisfies the n equations

$$\frac{\partial L}{\partial x_j} = \frac{\partial f}{\partial x_j} - \sum_{i=1}^m \lambda_i \frac{\partial g^i(\mathbf{x})}{\partial x_j} = 0 \quad \text{for } j = 1, \ldots, n. \qquad (7.3.5)$$

The maximizing point $\hat{\mathbf{x}}$ depends on $\boldsymbol{\lambda}$. For a different $\boldsymbol{\lambda}$ in that δ-neighborhood of $\boldsymbol{\lambda}^\circ$, we may have a different $\hat{\mathbf{x}}$. Hence the maximum of $L(\mathbf{x}, \boldsymbol{\lambda})$ with respect to $\mathbf{x}$ is a function of $\boldsymbol{\lambda}$, and we can call it $h(\boldsymbol{\lambda})$:

$$\max_{\mathbf{x}} L(\mathbf{x}, \boldsymbol{\lambda}) = L(\hat{\mathbf{x}}, \boldsymbol{\lambda}) = h(\boldsymbol{\lambda}). \qquad (7.3.6)$$

Of course, if $\boldsymbol{\lambda} = \boldsymbol{\lambda}^\circ$, then $\hat{\mathbf{x}} = \mathbf{x}^\circ$ and

$$\max_{\mathbf{x}} L(\mathbf{x}, \boldsymbol{\lambda}^\circ) = L(\mathbf{x}^\circ, \boldsymbol{\lambda}^\circ) = h(\boldsymbol{\lambda}^\circ).$$

Moreover, if $\mathbf{x}$ satisfies the constraints $g^i(\mathbf{x}) = b_i$, so that $L(\mathbf{x}, \boldsymbol{\lambda})$ coincides with $f(\mathbf{x})$, then

$$\max_{\substack{\mathbf{x} \text{ satisfying} \\ \text{constraints}}} L(\mathbf{x}, \boldsymbol{\lambda}) = \max_{\mathbf{x}} f(\mathbf{x}) = f(\mathbf{x}^\circ) = L(\mathbf{x}^\circ, \boldsymbol{\lambda}^\circ) = h(\boldsymbol{\lambda}^\circ). \qquad (7.3.7)$$

To see that the function $L(\mathbf{x}, \boldsymbol{\lambda})$ has a minimum with respect to $\boldsymbol{\lambda}$ at $\boldsymbol{\lambda}^\circ$, notice that for those $\mathbf{x}$ and $\boldsymbol{\lambda}$ which are related in the solution of the

---

[6] It is degenerate in that we have strict equality on the right hand side of (7.3.3).

equations in (7.3.5),

$$L(\mathbf{x}, \lambda) = h(\lambda).$$

In (7.3.7), the range of variations of $\mathbf{x}$ is limited to those which satisfy the constraints $g^i(\mathbf{x}) = b_i$, whereas in (7.3.6) the $\mathbf{x}$'s are unconstrained. So we must have

$$h(\lambda^\circ) \leqslant h(\lambda). \qquad (7.3.8)$$

Combining (7.3.4) with (7.3.7) and (7.3.8), we get

$$L(\mathbf{x}, \lambda^\circ) \leqslant L(\mathbf{x}^\circ, \lambda^\circ) = h(\lambda^\circ) \leqslant h(\lambda).$$

However, even if $\lambda \neq \lambda^\circ$, as long as $\mathbf{x}$ satisfies the constraints $g^i(\mathbf{x}) = b_i$, $L(\mathbf{x}^\circ, \lambda) = f(\mathbf{x}^\circ)$. And, since $f(\mathbf{x}^\circ) = L(\mathbf{x}^\circ, \lambda^\circ)$, we have the degenerate saddle point

$$L(\mathbf{x}, \lambda^\circ) \leqslant L(\mathbf{x}^\circ, \lambda^\circ) = L(\mathbf{x}^\circ, \lambda).$$

Although the above discussion does not constitute a proof, it shows that the K–T conditions for $(\mathbf{x}^\circ, \lambda^\circ)$ to be the maximum for $f(\mathbf{x})$ of the nonlinear programming problem are the same necessary conditions for the Lagrangian function $L(\mathbf{x}, \lambda)$ to have a saddle point at $(\mathbf{x}^\circ, \lambda^\circ)$ for $\mathbf{x} \geqslant 0$ and λ with the correct signs according to (7.2.3).

7.3.3 The K–T conditions and duality

As we just saw, the problem of finding the minimum of $L(\mathbf{x}, \lambda)$ with respect to λ, subject to the K–T conditions, is a dual to the problem of finding its maximum with respect to $\mathbf{x}$ in a programming problem. There are several theorems, using different assumptions, which deal with dual problems in nonlinear programming. In particular, the reader may be interested in Dantzig and Cottle (1967) and Wolfe (1961). We shall, however, direct our attention to the K–T conditions and duality in linear programming, which, after all, is a special case of the general programming problem.

Our λ_i° here correspond to the y_i° of Chapter 6. In linear programming, the functions $f = \sum_{j=1}^m c_j x_j$ and $g^i = \sum_{j=1}^m a_{ij} x_j$, being linear, are both concave and convex, so that the K–T conditions are sufficient as well as necessary for $(\mathbf{x}^\circ, \lambda^\circ)$ to be a saddle point of the Lagrangian function

$$L(x, \lambda) = f(x) + \sum_{i=1}^m \lambda_i \left[b_i - \sum_{j=1}^n a_{ij} x_j \right]$$

when $\mathbf{x} \geqslant 0$ and the components λ_i of λ have the right sign (depending on whether $\sum_{j=1}^n a_{ij} x_j \leqslant$ or $\geqslant b_i$), and for $(\mathbf{x}^\circ, \lambda^\circ)$ to be an optimal solution to the dual linear problems:
Maximize

$$z_{\mathbf{x}} = \sum_{j=1}^n c_j x_j, \quad \text{subject to} \quad \sum_{j=1}^n a_{ij} x_j \leqslant b_i \quad \text{and} \quad x_j \geqslant 0,$$

and
Minimize

$$z_\lambda = \sum_{i=1}^{m} \lambda_i b_i, \quad \text{subject to} \quad \sum_{i=1}^{m} a_{ij}\lambda_i \geqslant c_j, \qquad i=1,\ldots,m; j=1,\ldots,n.$$

Moreover, recall that

$$\sum_j a_{ij}x_j \pm x_{si} = b_i, \quad i.e., \quad \pm x_{si} = b_i - \sum_j a_{ij}x_j, \quad \text{for } i=1,\ldots,m,$$

and that at the optimum point (x°, λ°),

$$\lambda_i^\circ \cdot x_{si}^\circ = \lambda_i^\circ \left(b_i - \sum_j a_{ij}x_j^\circ \right) = 0.$$

Hence we can write $\sum_j a_{ij}x_j^\circ$ as b_i without affecting the value of $L(x^\circ, \lambda^\circ)$. Also,

$$\partial f(x^\circ)/\partial x_j = c_j \quad \text{and} \quad \partial g^i(x^\circ)/\partial x_j = a_{ij}.$$

Therefore the K–T condition given by (1') in (7.2.5),

$$\sum_{j=1}^{n} x_j^\circ \left[\partial f(x^\circ)/\partial x_j - \sum_{i=1}^{m} \lambda_i^\circ a_{ij} \right] = 0,$$

can be written as

$$\sum_{j=1}^{n} x_j^\circ \frac{\partial f(x^\circ)}{\partial x_j} = \sum_{i=1}^{m} \lambda_i^\circ \left(\sum_{j=1}^{n} a_{ij}x_j^\circ \right) = \sum_{i=1}^{m} \lambda_i^\circ b_i$$

or

$$\sum_{j=1}^{n} c_j x_j^\circ = \sum_{i=1}^{m} b_i \lambda_i^\circ.$$

This is precisely the Duality theorem (Theorem 6.1.6), which states that

$$z_x^\circ = z_y^\circ,$$

with y_i° of Chapter 6 replaced by λ_i° of this chapter.

Furthermore, the K–T condition given in (7.2.2),

$$\frac{\partial f(x^\circ)}{\partial x_j} - \sum_i \lambda_i^\circ \frac{\partial g^i(x^\circ)}{\partial x_j} \leqslant 0,$$

translated into linear programming terminology reads

$$c_j \leqslant \sum_{i=1}^{m} \lambda_i^\circ a_{ij},$$

which compares with the dual inequality constraints of (6.1.2), with λ° being the optimal solution to the dual linear programming problem:
Minimize

$$z_\lambda = \sum_{i=1}^{m} \lambda_i b_i$$

subject to

$$\sum_{i=1}^{m} a_{ij}\lambda_i \geqslant c_j, \qquad j=1,\ldots,n$$

$$\lambda_i \geqslant 0, \qquad i=1,\ldots,u,$$

$$\lambda_i \leqslant 0, \qquad i=u+1,\ldots,v,$$

$$\lambda_i \text{ unrestricted}, \qquad i=v+1,\ldots,m.$$

Thus the K–T theorem establishes the relationship between the primal and dual linear programming problems, and the optimal solution of the primal is the same as the optimal solution of the dual. Indeed, the λ_i of the K–T conditions form a bridge between the classical Lagrange multipliers and the optimal dual variables $y_i^{\,o}$ of linear programming. Therefore these λ_i admit the same interpretation as imputed values in economic analysis.

In passing, it is of interest to comment that since

$$f(\mathbf{x}^o) = h(\boldsymbol{\lambda}^o) = \min_{\boldsymbol{\lambda}} L(\mathbf{x},\boldsymbol{\lambda}),$$

and

$$\max_{\mathbf{x}} L(\mathbf{x},\boldsymbol{\lambda}) = h(\boldsymbol{\lambda}),$$

$$\min_{\boldsymbol{\lambda}} h(\boldsymbol{\lambda}) = h(\boldsymbol{\lambda}^o),$$

it follows that

$$f(\mathbf{x}^o) = \min_{\boldsymbol{\lambda}} \max_{\mathbf{x}} L(\mathbf{x},\boldsymbol{\lambda}),$$

which shows that $f(\mathbf{x}^o)$ is the solution to a *minimax* problem. This is one of the notions in game theory.

7.4 Quadratic programming

The K–T conditions, important as they are, do not provide us with a computational technique for finding an optimal solution to a nonlinear programming problem as the simplex algorithm does for linear programming, except in some simple cases and in solving some types of quadratic programming problems.

The term "quadratic programming" as it appears in popular usage refers to a problem with linear constraints and an objective function which is the sum of a linear and a quadratic form. That is, the objective function is of the form

$$z = f(\mathbf{x}) = \sum_{j=1}^{n} c_j x_j - \frac{1}{2} \sum_{k=1}^{n} \sum_{j=1}^{n} q_{jk} x_j x_k,$$

where $q_{jk} = q_{kj}$ and the constraints are of the form

$$g^i(\mathbf{x}) = \sum_{j=1}^{n} a_{ij} x_j \ (\leqslant, =, \text{or} \geqslant) \ b_i, \qquad i = 1, \ldots, m,$$

$$x_j \geqslant 0, \qquad j = 1, \ldots, n.$$

We do not intend to go into the subject in depth but will restrict ourselves to an approach which is based on the K–T conditions and uses the simplex method of computation. Needless to say, the subject of quadratic programming is much too broad to be covered in one short section. There are various computational techniques, but we will restrict the discussion to the algorithm due to Wolfe (1959). For our purposes, we shall assume that the quadratic form is positive definite, that is,

$$\sum_{k=1}^{n} \sum_{j=1}^{n} q_{jk} x_j x_k > 0,$$

for all values of x_j except if $x_1 = x_2 = \ldots = x_n = 0$.[7] It can be proved that if the quadratic form is positive definite or semidefinite, the objective function is concave. Since the constraints are linear, the set of feasible solutions is convex. Therefore the K–T conditions will be not only necessary but also sufficient for an optimum solution. For the sake of convenience, we shall consider only maximization problems and we shall take the linear constraint inequalities to be in the direction $\leqslant b_i$ where b_i will be assumed to be positive, $i = 1, \ldots, m$. This should not affect the generality of our discussion in any way because, as we know, a minimization problem can be converted into a maximization problem by maximizing the negative of the objective function. As for the directions of the inequalities, the K–T condition given by (7.2.3) is intended to take care of the matter by requiring λ to be either positive, negative, or unrestricted in sign. In order to simplify our notation, we shall omit the superscript "o" when indicating the values of $\mathbf{x}$ or λ which correspond to the optimal solution, unless its inclusion is made necessary for reasons of emphasis.

First, we add slack variables x_{si}, $i = 1, \ldots, m$, to the inequalities so that the constraints now become

$$\sum a_{ij} x_j + x_{si} = b_i \quad \text{for } i = 1, \ldots, m.$$

---

[7] Actually, Wolfe's method assumes only positive semidefiniteness. However, in this case, there is a possibility that the quadratic programming problem may have an unbounded solution or that, if it does have a finite optimal solution, this optimal solution is not obtained when the computational procedure terminates. If the linear form is 0, that is, if $c_1 = c_2 \ldots = c_n = 0$, or if the quadratic form is positive definite, then we will not need to worry. One can use a perturbation technique (see Barankin and Dorfman (1958), pp. 299–306) to achieve positive definiteness, or a parametric technique (see Wolfe (1959)) to ensure that the computations will terminate with an unbounded optimal solution.

According to the K–T conditions given by (7.2.2), there should be numbers $\lambda_i \geq 0$, for $i = 1, \ldots, m$, such that

$$c_j - \sum_{k=1}^{n} q_{jk} x_k - \sum_{i=1}^{m} a_{ij} \lambda_i \leq 0 \quad \text{for } j = 1, \ldots, n.$$

If we let u_j be the negative of the above expression, then $u_j \geq 0$, that is,

$$u_j = \sum_{k=1}^{n} q_{jk} x_k + \sum_{i=1}^{m} a_{ij} \lambda_i - c_j \geq 0 \quad \text{for } j = 1, \ldots, n.$$

We now have the K–T conditions for our quadratic programming problem:

$$\sum_{k=1}^{n} q_{jk} x_k + \sum_{i=1}^{m} a_{ij} \lambda_i - u_j = c_j, \qquad j = 1, \ldots, n; \tag{7.4.1}$$

$$\sum_{j=1}^{n} a_{ij} x_j + x_{si} = b_i, \qquad i = 1, \ldots, m; \tag{7.4.2}$$

$$x_j u_j = 0, \qquad j = 1, \ldots, n; \tag{7.4.3}$$

$$x_{si} \lambda_i = 0, \qquad i = 1, \ldots, m; \tag{7.4.4}$$

$$x_j \geq 0, \quad u_j \geq 0, \quad \lambda_i \geq 0, \quad x_{si} \geq 0, \qquad j = 1, \ldots, n; \ i = 1, \ldots, m.$$

Since the K–T conditions are both necessary and sufficient, we know that $\mathbf{x} = (x_1, \ldots, x_n)^{\mathrm{T}}$ is an optimal solution to the quadratic programming problem if we can find $\lambda_i \geq 0$ and $u_j \geq 0$ ($i = 1, \ldots, m$ and $j = 1, \ldots, n$) such that the above equations are satisfied. The $n + m$ equations in (7.4.1) and (7.4.2) are linear constraints involving $2(n + m)$ variables (the n x_j's, n u_j's, m λ_i's, and m x_{si}'s). Of these, at most $n + m$ are positive, because (7.4.3) and (7.4.4) say that if k of the x_j's (or k of the λ_i's) are positive, then at least k of the u_j's (or k of the x_{si}'s) must be zero. These $n + m$ variables can be a basic solution to the system of linear equations given by (7.4.1) and (7.4.2), and we may use the simplex technique of linear programming to find the basic (nondegenerate) feasible solutions, with the important restriction that x_j and u_j cannot both be basic, nor x_{si} and λ_i both be basic at the same time.[8]

In considering an initial basic feasible solution, we may use x_{si}, $i = 1, \ldots, m$, as the initial basic variables for (7.4.2) because we have assumed each b_i to be positive.[9] On the other hand, the c_j's in most practical

---

[8] Notice that, at this stage, we are not attempting to solve any linear programming problem. We are only going to use the simplex technique to find a basic feasible solution to the system of $n + m$ equations given by (7.4.1) and (7.4.2).

[9] Again, this assumption of positivity is only for convenience. If some b_i were originally < 0, we could make some simple adjustments. See Subsections 5.4.2 and 5.4.3.

problems are positive and we cannot use the u_j's in (7.4.1) as the initial basic feasible variables. When the initial basic feasible variables are not obvious, we resort to the technique of introducing nonnegative artificial variables $v_j, j=1,\ldots,n$, and driving them to zero. This method is described in Subsection 5.4.1. Since some of the c_j's may be negative, we make the following modifications: change (7.4.1) to (7.4.1'),

$$\sum_{k=1}^{n} q_{jk}x_k + \sum_{i=1}^{m} a_{ij}\lambda_i - u_j + v_{j1} - v_{j2} = c_j, \qquad j=1,\ldots,m;$$

set all the x_k, λ_i, and u_j equal to 0; set $v_{j1}=c_j$ and $v_{j2}=0$ if $c_j \geq 0$, but set $v_{j1}=0$ and $v_{j2}=c_j$ if $c_j<0$. Or, to further simplify our notation, let the artificial variable $v_j=1$ for $j=1,\ldots,n$, and write (7.4.1') as

$$\sum_{k=1}^{n} q_{jk}x_k + \sum_{i=1}^{m} a_{ij}\lambda_i - u_j + c_j v_j = c_j, \qquad j=1,\ldots,n. \qquad (7.4.1')$$

Notice that (7.4.1') becomes (7.4.1) when the v_j's are driven to 0. Our initial basic feasible solution is $x_{si}=b_i$, where $i=1,\ldots,m$, and $v_j=1, j=1,\ldots,n$, and all the other variables equal 0. In order that the feasible solution to the artificial problem is also feasible for the original problem, the v_j's must be driven to 0.

To summarize the procedure, we use the initial basic feasible solution given above and apply the simplex method to minimize $v=\sum_{j=1}^{n}v_j$ subject to:

$$\sum_{k=1}^{n} q_{jk}x_k + \sum_{i=1}^{m} a_{ij}\lambda_i - u_j + c_j v_j = c_j, \qquad j=1,\ldots,n,$$

$$\sum_{j=1}^{n} a_{ij}x_j + x_{si} = b_i \qquad i=1,\ldots,m,$$

$$x_j \geq 0, \qquad u_j \geq 0, \qquad v_j \geq 0, \qquad \lambda_i \geq 0, \qquad x_{si} \geq 0, \qquad j=1,\ldots,n; i=1,\ldots,m.$$

When carrying out the simplex algorithm, make sure that x_j does not become a basic variable when u_j already is one (and *vice versa*) and, similarly, that λ_i (or x_{si}) does not become a basic variable if x_{si} (or λ_i) already is one. The question remains whether the simplex iterations, when subject to the constraints $x_j u_j=0$ and $x_{si}\lambda_i=0$, can be carried to the desired result that

$$v= \sum_{j=1}^{n} v_j=0.$$

It can be proved that the iterations will terminate with $v=0$ or $c_j=0$ for $j=1,\ldots,n$ if the quadratic form $\sum_{j=1}^{n}\sum_{k=1}^{n}q_{jk}x_jx_k$ is positive definite, and methods have been devised to take care of the semidefinite case (see Footnote 7 earlier in this section). The solution to this revised linear programming problem is also the optimal solution to our original quadratic programming problem, and we find that the computations are approximately those for a linear programming problem with $m+n$ constraints.

Example 1. Let us maximize

$$z = 8x_1 + 10x_2 - (x_1^2 + x_2^2)$$

subject to

$$3x_1 + 2x_2 \leqslant 6, \qquad x_1 \geqslant 0, \qquad x_2 \geqslant 0.$$

Here $c_1 = 8$, $c_2 = 10$, $q_{11} = 2$, $q_{12} = q_{21} = 0$, $q_{22} = 2$, $n = 2$, $a_{11} = 3$, $a_{12} = 2$, $b = 6$, with $m = 1$. The quadratic form is clearly positive definite. The K–T conditions require that

$$
\begin{array}{llllll}
3x_1 + 2x_2 + x_s & & & & = 6, \\
2x_1 & + 3\lambda - u_1 & + 8v_1 & & = 8, & \qquad (7.4.5) \\
& 2x_2 & + 2\lambda & - u_2 & + 10v_2 = 10,
\end{array}
$$

and we now wish to minimize

$$v = v_1 + v_2$$

subject to (7.4.5) and the nonnegativity requirements for all the variables. This is the same as maximizing $w = -v$, that is, maximizing

$$w = -v_1 - v_2 = -2 + \tfrac{1}{4}x_1 + \tfrac{1}{5}x_2 + \tfrac{23}{40}\lambda - \tfrac{1}{8}u_1 - \tfrac{1}{10}u_2,$$

by solving for v_1 and v_2 in terms of the other variables in (7.4.5). The initial basic feasible solution is $x_s = 6$, $v_1 = 1$, $v_2 = 1$. We shall carry out the simplex algorithm and write our tableau in exactly the same form as we did in Chapter 5. When we look at the objective equation

$$w - \tfrac{1}{4}x_1 - \tfrac{1}{5}x_2 - \tfrac{23}{40}\lambda + \tfrac{1}{8}u_1 + \tfrac{1}{10}u_2 = -2,$$

we may be tempted to enter λ as the basic variable. However, a little calculation will quickly show that in this case x_s will not be the leaving basic variable and we will be violating the requirement that λ and x_s cannot both be basic at the same time. Therefore, our next best choice for the entering basic variable is x_1, which will replace x_s. The calculations are listed in the tableau below.

The solution $x_1 = 4/13$, $x_2 = 33/13$ is also the optimal solution to our original problem.

Example 2. The following illustrates the advantages in making a judicious choice of the entering and leaving basic variables. We wish to maximize

$$z = 4x_1 + x_2 - x_1^2$$

subject to

$$2x_1 + x_2 \leqslant 5,$$

$$2x_1 - x_2 \leqslant 3,$$

$$x_1 \geqslant 0, \qquad x_2 \geqslant 0.$$

Table 7.4.1

Iteration	Basis	x_1	x_2	x_s	λ	u_1	u_2	v_1	v_2	Current value	Ratio
1	w	$-\frac{1}{4}$	$-\frac{1}{5}$	0	$-\frac{23}{40}$	$\frac{1}{8}$	$\frac{1}{10}$	0	0	-2	
	x_s	3	2	1	0	0	0	0	0	6	2 (min)
	v_1	$\frac{1}{4}$	0	0	$\frac{3}{8}$	$-\frac{1}{8}$	0	1	0	1	4
	v_2	0	$\frac{1}{5}$	0	$\frac{1}{5}$	0	$-\frac{1}{10}$	0	1	1	
2	w	0	$-\frac{1}{30}$	$\frac{1}{12}$	$-\frac{23}{40}$	$\frac{1}{8}$	$\frac{1}{10}$	0	0	$-\frac{3}{2}$	
	x_1	1	$\frac{2}{3}$	$\frac{1}{3}$	0	0	0	0	0	2	
	v_1	0	$-\frac{1}{6}$	$-\frac{1}{12}$	$\frac{3}{8}$	$-\frac{1}{8}$	0	1	0	$\frac{1}{2}$	$\frac{4}{3}$ (min)
	v_2	0	$\frac{1}{5}$	0	$\frac{1}{5}$	0	$-\frac{1}{10}$	0	1	1	5
3	w	0	$-\frac{13}{45}$	$-\frac{2}{45}$	0	$-\frac{1}{15}$	$\frac{1}{10}$	$\frac{23}{15}$	0	$-\frac{11}{15}$	
	x_1	1	$\frac{2}{3}$	$\frac{1}{3}$	0	0	0	0	0	2	3
	λ	0	$-\frac{4}{9}$	$-\frac{2}{9}$	1	$-\frac{1}{3}$	0	$\frac{8}{3}$	0	$\frac{4}{3}$	
	v_2	0	$\frac{13}{45}$	$\frac{2}{45}$	0	$\frac{1}{15}$	$-\frac{1}{10}$	$-\frac{8}{15}$	1	$\frac{11}{15}$	$\frac{33}{13}$ (min)
4	w	0	0	0	0	0	0	$\frac{1}{3}$	1	0	
	x_1	1	0	$\frac{3}{13}$	0	$-\frac{2}{13}$	$\frac{3}{13}$	$\frac{48}{39}$	$-\frac{30}{13}$	$\frac{4}{13}$	
	λ	0	0	$-\frac{2}{13}$	1	$-\frac{3}{13}$	$-\frac{2}{13}$	$-\frac{24}{13}$	$\frac{20}{13}$	$\frac{32}{13}$	
	x_2	0	1	$\frac{2}{13}$	0	$\frac{3}{13}$	$-\frac{9}{26}$	$-\frac{24}{13}$	$\frac{45}{13}$	$\frac{33}{13}$	

This is a simple problem which can be solved graphically. The objective function is a family of parabolas. Looking at the graph, we see that the highest curve which will satisfy the constraints is the one which intersects the line $2x_1 + x_2 = 5$. At the point of intersection the slope of the curve is the same as the slope of the line, namely -2. Solving the equations

$$\frac{dx_2}{dx_1} = 2x_1 - 4 = -2 \quad \text{and} \quad 2x_1 + x_2 = 5$$

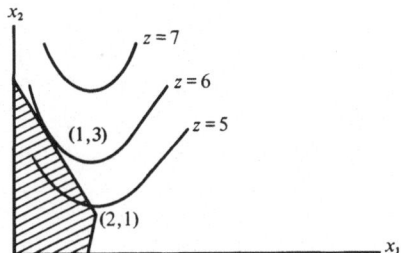

Figure 7.4.1.

together we get $x_1=1$, $x_2=3$ as the solution to our problem, yielding $z=6$.
Let us try using the technique described in this section. We find

$$c_1=4, \qquad c_2=1, \qquad q_{11}=2, \qquad q_{12}=q_{21}=q_{22}=0,$$
$$a_{11}=2, \qquad a_{12}=1, \qquad a_{21}=2, \qquad a_{22}=-1, \qquad b_1=5, \qquad b_2=3,$$
$$m=2, \qquad n=2.$$

The quadratic form is positive semidefinite. From the K–T conditions, we
have

$$2x_1+x_2+x_{s1} \qquad\qquad\qquad\qquad\qquad\qquad =5,$$
$$2x_1-x_2 \qquad +x_{s2} \qquad\qquad\qquad\qquad\qquad =3,$$
$$2x_1 \qquad\qquad\qquad +2\lambda_1+2\lambda_2-u_1 \qquad +4v_1 \qquad =4,$$
$$\lambda_1-\lambda_2 \qquad -u_2 \qquad +v_2=1,$$

and we wish to maximize

$$w=-v_1-v_2=\frac{1}{2}x_1+\frac{3}{2}\lambda_1-\frac{1}{2}\lambda_2-\frac{1}{4}u_1-u_2-2$$

subject to the above equations and nonnegativity conditions.
   At Iteration 1, x_1, instead of λ_1, is chosen as the entering basic variable
because x_{s1} is already a basic variable and will not be leaving. At Iteration
2, x_{s1} and v_1 tie as the leaving basic variable. If we choose v_1, we would be
at Iteration 3′ and will find that we cannot proceed any further, because λ_1

Table 7.4.2

Iteration	Basis	x_1	x_2	x_{s1}	x_{s2}	λ_1	λ_2	u_1	u_2	v_1	v_2	Current value	Ratio
	w	$-\frac{1}{2}$	0	0	0	$-\frac{3}{2}$	$\frac{1}{2}$	$\frac{1}{4}$	1	0	0	-2	
	x_{s1}	2	1	1	0	0	0	0	0	0	0	5	$\frac{5}{2}$
1	x_{s2}	2	-1	0	1	0	0	0	0	0	0	3	$\frac{3}{2}$ (min)
	v_1	$\frac{1}{2}$	0	0	0	$\frac{1}{2}$	$\frac{1}{2}$	$-\frac{1}{4}$	0	1	0	1	2
	v_2	0	0	0	0	1	-1	0	-1	0	1	1	
	w	0	$-\frac{1}{4}$	0	$\frac{1}{4}$	$-\frac{3}{2}$	$\frac{1}{2}$	$\frac{1}{4}$	1	0	0	$-\frac{5}{4}$	
	x_{s1}	0	2	1	-1	0	0	0	0	0	0	2	1
2	x_1	1	$-\frac{1}{2}$	0	$\frac{1}{2}$	0	0	0	0	0	0	$\frac{3}{2}$	
	v_1	0	$\frac{1}{4}$	0	$-\frac{1}{4}$	$\frac{1}{2}$	$\frac{1}{2}$	$-\frac{1}{4}$	0	1	0	$\frac{1}{4}$	1
	v_2	0	0	0	0	1	-1	0	-1	0	1	1	
	w	0	0	0	0	-1	1	0	1	1	0	-1	
	x_{s1}	0	0	1	1	-4	-4	2	0	-8	0	0	
3′	x_1	1	0	0	0	1	1	$-\frac{1}{2}$	0	2	0	2	2
	x_2	0	1	0	-1	2	2	-1	0	4	0	1	$\frac{1}{2}$
	v_2	0	0	0	0	1	-1	0	-1	0	1	1	1

Table 7.4.3.

Iteration	Basis	x_1	x_2	x_{s1}	x_{s2}	λ_1	λ_2	u_1	u_2	v_1	v_2	Current value	ratio
	w	0	0	$\frac{1}{8}$	$\frac{1}{8}$	$-\frac{3}{2}$	$\frac{1}{2}$	$\frac{1}{4}$	1	0	0	-1	
	x_2	0	1	$\frac{1}{2}$	$-\frac{1}{2}$	0	0	0	0	0	0	1	
3	x_1	1	0	$\frac{1}{4}$	$\frac{1}{4}$	0	0	0	0	0	0	2	
	v_1	0	0	$-\frac{1}{8}$	$-\frac{1}{8}$	$\frac{1}{2}$	$-\frac{1}{2}$	$-\frac{1}{4}$	0	1	0	0	0
	v_2	0	0	0	0	1	-1	0	-1	0	1	1	1
	w	0	0	$-\frac{1}{4}$	$-\frac{1}{4}$	0	2	$-\frac{1}{2}$	1	3	0	-1	
	x_2	0	1	$\frac{1}{2}$	$-\frac{1}{2}$	0	0	0	0	0	0	1	
4	x_1	1	0	$\frac{1}{4}$	$\frac{1}{4}$	0	0	0	0	0	0	2	8
	λ_1	0	0	$-\frac{1}{4}$	$-\frac{1}{4}$	1	1	$-\frac{1}{2}$	0	2	0	0	
	v_2	0	0	$\frac{1}{4}$	$\frac{1}{4}$	0	-2	$\frac{1}{2}$	-1	-2	1	1	4
	x_{s1}	0	0	0	0	0	0	0	0	1	1	0	
	x_2	0	1	1	0	0	-4	1	-2	-4	2	3	
5	x_1	1	0	0	0	0	2	$-\frac{1}{2}$	1	2	-1	1	
	λ_1	0	0	0	0	1	-1	0	-1	0	1	1	
	x_{s2}	0	0	1	1	0	-8	2	-4	-8	4	4	

cannot be entered with x_{s1} still a basic variable. So we return to Iteration 2 and choose x_{s1} as the leaving variable. The new Iteration 3 and following iterations are shown in Table 7.4.3.

The solution is $x_1 = 1$, $x_2 = 3$, $z = 6$. It may be of interest to call attention to the values $\lambda_1 = 1$, $\lambda_2 = 0$, $x_{s1} = 0$, and $x_{s2} = 4$. This result agrees with the graph, which shows that the second constraint

$$2x_1 - x_2 \leqslant 3$$

is not active at the optimum point, and, therefore, $\lambda_2 = 0$. The first constraint, however, is active; thus $x_{s1} = 0$ and $\lambda_1 = 1$.

Suppose one were to forget that x_j and u_j cannot both be basic for the same j and decide at Iteration 4 to enter u_1 as the basic variable since it has the most negative coefficient in the objective equation. We would have Iterations 4′ and 5′ as shown in Table 7.4.4 and reach the incorrect conclusion that $x_1 = 2$, $x_2 = 1$, $z = 5$ as the optimal solution.

In Section 7.3 we pointed out that finding the minimum of

$$L(\mathbf{x}, \boldsymbol{\lambda}) = f(\mathbf{x}) + \sum_{i=1}^{m} \lambda_i \left[b_i - \sum_{j=1}^{n} a_{ij} x_j \right]$$

with respect to $\boldsymbol{\lambda}$ is the dual of the linear programming problem of finding

Table 7.4.4.

Iteration	Basis	Current coefficient										Current value	Ratio
		x_1	x_2	x_{s1}	x_{s2}	λ_1	λ_2	u_1	u_2	v_1	v_2		
	w	0	0	$-\frac{1}{4}$	$-\frac{1}{4}$	0	2	$-\frac{1}{2}$	1	3	0	-1	$-\frac{1}{2}$
	x_2	0	1	$\frac{1}{2}$	$-\frac{1}{2}$	0	0	0	0	0	0	1	
4'	x_1	1	0	$\frac{1}{4}$	$\frac{1}{4}$	0	0	0	0	0	0	2	
	λ_1	0	0	$-\frac{1}{4}$	$-\frac{1}{4}$	1	1	$-\frac{1}{2}$	0	2	0	0	
	v_2	0	0	$\frac{1}{4}$	$\frac{1}{4}$	0	-2	$\frac{1}{2}$	-1	-2	1	1	2
	w	0	0	0	0	0	0	0	0	1	1	0	
	x_2	0	1	$\frac{1}{2}$	$-\frac{1}{2}$	0	0	0	0	0	0	1	
5'	x_1	1	0	$\frac{1}{4}$	$\frac{1}{4}$	0	0	0	0	0	0	2	
	λ_1	0	0	0	0	1	-1	0	-1	0	1	1	
	u_1	0	0	$\frac{1}{2}$	$\frac{1}{2}$	0	-4	1	-2	-4	2	2	

the maximum of

$$f(\mathbf{x}) = \sum_{j=1}^{n} c_j x_j$$

subject to

$$\sum_{j=1}^{n} a_{ij} x_j \leqslant b_i, \qquad x_j \geqslant 0, \qquad j = 1, \ldots, n; \, i = 1, \ldots, m.$$

Let us take a very brief look at the dual of the quadratic programming problem of finding the maximum of

$$z = \sum_{j=1}^{n} c_j x_j - \frac{1}{2} \sum_{j=1}^{n} \sum_{k=1}^{n} q_{jk} x_j x_k$$

subject to

$$\sum_{j=1}^{n} a_{ij} x_j \leqslant b_i, \qquad i = 1, \ldots, m,$$

$$x_j \geqslant 0, \qquad b_i \geqslant 0, \qquad j = 1, \ldots, n; \, i = 1, \ldots, m.$$

The dual to this should be to minimize

$$w = \sum_{j=1}^{n} c_j x_j - \frac{1}{2} \sum_{j=1}^{n} \sum_{k=1}^{n} q_{jk} x_j x_k + \sum_{i=1}^{m} \lambda_i \left[b_i - \sum_{j=1}^{n} a_{ij} x_j \right], \qquad (7.4.6)$$

subject to

$$\sum_{k=1}^{n} q_{jk} x_k + \sum_{i=1}^{m} a_{ij} \lambda_i \geqslant c_j, \qquad j = 1, \ldots, n,$$

$$x_j \geqslant 0, \qquad \lambda_i \geqslant 0, \qquad j = 1, \ldots, n; \, i = 1, \ldots, m. \qquad (7.4.7)$$

Notice that (7.4.7) is simply the K–T condition given by (7.2.2) requiring

$$\frac{\partial f}{\partial x_j} - \sum_{i=1}^{m} \lambda_i \frac{g^i(\mathbf{x}^{\circ})}{\partial x_j} \leq 0.$$

However, we wish to have min w correspond with the optimal solution to the primal problem. By the K–T condition (7.2.2), in order for any $\mathbf{x} \geq 0$ and $\boldsymbol{\lambda}$ to be an optimal solution to the primal problem,

$$x_j\left(c_j - \sum_{k=1}^{n} q_{jk}x_k - \sum_{i=1}^{m} a_{ij}\lambda_i\right) = 0, \qquad j=1,\ldots,n,$$

or, upon summing over j,

$$\sum_{j=1}^{n} c_j x_j = \sum_{j=1}^{n}\sum_{k=1}^{n} q_{jk}x_j x_k + \sum_{j=1}^{n}\sum_{i=1}^{m} a_{ij}\lambda_i x_j. \qquad (7.4.8)$$

Substitute (7.4.8) into (7.4.6), and the dual problem becomes:
Minimize

$$w = \frac{1}{2}\sum_{j=1}^{n}\sum_{k=1}^{n} q_{jk}x_j x_k + \sum_{i=1}^{m} \lambda_i b_i, \qquad (7.4.9)$$

subject to (7.4.7).

To see that $z \leq w$ (compare with Theorem 6.1.2) and max $z = $ min w (compare with Theorem 6.1.6), notice first that

$$w - z = \sum_{i=1}^{m} \lambda_i\left[b_i - \sum_{j=1}^{n} a_{ij}x_j\right] \geq 0.$$

Next, consider the optimal solution $(\mathbf{x}^{\circ}, \boldsymbol{\lambda}^{\circ})$ to the primal problem. From the K–T condition given by (7.2.4),

$$\lambda_i^{\circ}\left[b_i - \sum_{j=1}^{n} a_{ij}x_j^{\circ}\right] = 0 \quad \text{for } i = 1,\ldots,m,$$

and from (7.4.8) we see that

$$\max z = z^{\circ} = \sum_{j=1}^{n} c_j x_j^{\circ} - \frac{1}{2}\sum_j\sum_k q_{jk}x_j^{\circ}x_k^{\circ} + \sum_{i=1}^{m} \lambda_i^{\circ}\left[b_i - \sum a_{ij}x_j^{\circ}\right]$$

$$= L(\mathbf{x}^{\circ}, \boldsymbol{\lambda}^{\circ}) = \frac{1}{2}\sum_j\sum_k q_{ji}x_j^{\circ}x_k^{\circ} + \sum_i \lambda_i^{\circ}b_i.$$

In general, for any $\mathbf{x} \geq 0$ and $\boldsymbol{\lambda}$ satisfying $c_j \leq \sum_k q_{jk}x_k + \sum_i a_{ij}\lambda_i$, it is true that $L(\mathbf{x}, \boldsymbol{\lambda}) \leq w$, because

$$L(\mathbf{x}, \boldsymbol{\lambda}) = \sum_j c_j x_j - \frac{1}{2}\sum_i\sum_j q_{jk}x_j x_k + \sum_i \lambda_i b_i - \sum_i\sum_j a_{ij}\lambda_i x_j$$

$$\leq \frac{1}{2}\sum\sum q_{jk}x_j x_k + \sum \lambda_i b_i = w.$$

Therefore,

$$\min w = L(\mathbf{x}^\circ, \boldsymbol{\lambda}^\circ)$$

and

$$\max z = \min w.$$

It can be proved that if the primal problem has an optimal solution, so does the dual, and *vice versa*. Notice that the optimal solution $\mathbf{x}^\circ$ to the primal is part of the optimal solution $(\mathbf{x}^\circ, \boldsymbol{\lambda}^\circ)$ to the dual.

7.5 Separable functions

Our nonlinear programming problem may happen to be of the following type:
Maximize

$$z = f(x_1, x_2, \ldots, x_n) = \sum_{j=1}^{n} f_j(x_j)$$

subject to   (7.5.1)

$$g^i(x_1, \ldots, x_n) = \sum_{j=1}^{n} g_j^i(x_j) \{\leqslant, = \text{or} \geqslant\} b_i, \qquad i = 1, \ldots, m,$$

$$x_j \geqslant 0, \qquad j = 1, \ldots, n,$$

where f_j (or g_j^i) is a function of the variable x_j alone. Notice that each function, objective or constraint, is itself separable into a sum of n functions, each being a function of only one variable x_j, $j = 12, \ldots, n$. We shall assume that the functions f and g^i are all continuous and differentiable. Each function can be approximated by a piecewise linear function, and the problem may then be approximated by a linear programming problem.

To illustrate how the idea works, let us consider a simple continuous function $h(y)$ of one variable y which is defined for $0 \leqslant y \leqslant a$. Suppose the function looks like the one in Figure 7.5.1. Select $r+1$ convenient points y_k, not necessarily spaced evenly, with $y_0 = 0$, $y_0 < y_1 < y_2 \ldots < y_r = a$. Calculate the value of the function $h_k = h(y_k)$ at each y_k, and join each pair of points (y_k, h_k) and (y_{k+1}, h_{k+1}) by a straight line. We then have a piecewise linear approximation of $h(y)$ which we shall call $\bar{h}(y)$. The approximation can be made closer by subdividing the interval $0 \leqslant y \leqslant a$ into finer segments.

There are various ways to express the approximating function $\bar{h}(y)$ analytically in terms of the original function $h(y)$. Let us briefly consider one such way.

Notice that at the break points, $h_k = \bar{h}_k$ for $k = 0, 1, \ldots, r$. If we select any point y which lies in some subinterval, say $y_k \leqslant y \leqslant y_{k+1}$, we know that

$$\frac{\bar{h}(y) - h_k}{y - y_k} = \frac{h_{k+1} - h_k}{y_{k+1} - y_k},$$

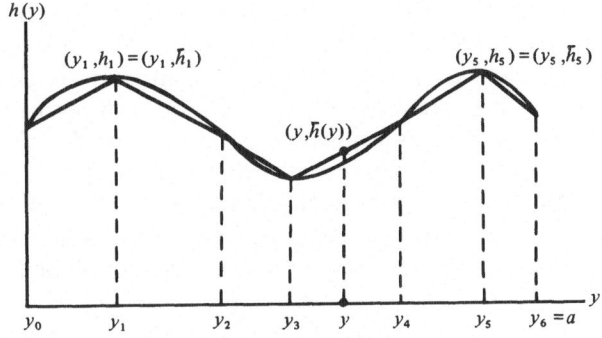

Figure 7.5.1.

that is,

$$\bar{h}(y) = h_k + \frac{h_{k+1} - h_k}{y_{k+1} - y_k}(y - y_k), \tag{7.5.2}$$

and that there exists a λ, with $0 \leqslant \lambda \leqslant 1$, such that

$$y = \lambda y_{k+1} + (1 - \lambda) y_k,$$

because y lies in the segment between y_k and y_{k+1}. Substituting, we see that

$$y - y_k = \lambda(y_{k+1} - y_k),$$

and that

$$\bar{h}(y) = h_k + \frac{h_{k+1} - h_k}{y_{k+1} - y_k}\lambda(y_{k+1} - y_k) = \lambda h_{k+1} + (1 - \lambda)h_k.$$

If we write λ as λ_{k+1} and $(1 - \lambda)$ as λ_k, then the above may be summarized as: For any y with $y_k \leqslant y \leqslant y_{k+1}$, there exists unique λ_k and λ_{k+1} such that

$$y = \lambda_k y_k + \lambda_{k+1} y_{k+1},$$

$$\bar{h}(y) = \lambda_k h_k + \lambda_{k+1} h_{k+1}, \tag{7.5.3}$$

$$\lambda_k + \lambda_{k+1} = 1, \qquad \lambda_k \geqslant 0, \qquad \lambda_{k+1} \geqslant 0.$$

Similarly, of course, if y lies in the segment between, say, y_{k+1} and y_{k+2}, we have

$$y = \lambda_{k+1} y_{k+1} + \lambda_{k+2} y_{k+2}, \qquad \bar{h}(y) = \lambda_{k+1} h_{k+1} + \lambda_{k+2} h_{k+2},$$

and

$$\lambda_{k+1} + \lambda_{k+2} = 1.$$

In fact, for any y in the interval $0 \leqslant y \leqslant a$, we can write

$$y = \sum_{k=0}^{r} \lambda_k y_k, \qquad \bar{h}(y) = \sum_{k=0}^{r} \lambda_k h_k,$$

$$\sum_{k=0}^{r} \lambda_k = 1, \qquad \lambda_k \geqslant 0, \qquad k = 0, 1, \ldots, r, \tag{7.5.4}$$

as long as we specify also that no more than two of the λ's may be positive, and that if two are positive, they must be adjacent. That is to say, the other λ's will all be zero, and only one λ or two adjacent ones are positive. When this requirement is observed, (7.5.4) is in effect no different from (7.5.3), because the awesome-looking expression $\Sigma_{k=0}^{r}\lambda_{k}y_{k}$ (or $\Sigma_{k=0}^{r}\lambda_{k}h_{k}$) actually contains at most two terms, the rest all being equal to 0.

We have shown how a continuous function of one variable may be approximated by a piecewise linear function of the same variable. Thus each continuous function $f_j(x_j)$ or $g_j^i(x_j)$ of a separable function

$$f(x_1, x_2, \ldots, x_n) = \sum_{j=1}^{n} f_j(x_j)$$

or

$$g^i(x_1, \ldots, x_n) = \sum_{j=1}^{n} g_j^i(x_j)$$

may be similarly approximated by $\bar{f_j}(x_j)$ or $\bar{g}_j^i(x_j)$, with $\bar{f_j}(x_j) = \Sigma_{k=0}^{r_j}\lambda_{jk}f_{jk}$, $\bar{g}_j^i(x_j) = \Sigma_{k=0}^{r_j}\lambda_{jk}g_{jk}^i$, in which we use the notation

$$f_{jk} = f_j(x_{jk}) \quad \text{and} \quad g_{jk}^i = g_j^i(x_{jk}),$$

where the x_{jk} are the break points for x_j in the interval $0 \le x_j \le a$ if we assume from physical considerations of the problem that each x_j would be bounded from above by a. That is,

$$x_j = \sum_{k=0}^{r_j} \lambda_{jk}x_{jk}, \qquad \sum_{k=0}^{r_j} \lambda_{jk} = 1, \qquad \lambda_{jk} \ge 0,$$

and for each given j only one or two adjacent λ_{jk} is (or are) positive.

A popular technique to solve a separable nonlinear program is to approximate each function by a piecewise linear one and treat the problem as a linear programming problem. For instance, the approximating problem of (7.5.1) would be:
Maximize

$$\bar{z} = \sum_{j=1}^{n} \bar{f_j}(x_j) = \sum_{j=1}^{n} \sum_{k=0}^{r_j} f_{jk}\lambda_{jk}$$

subject to

$$\bar{g}^i = \sum_{j=1}^{n} \bar{g}_j^i(x_j) = \sum_{j=1}^{n} \sum_{k=0}^{r_j} g_{jk}^i\lambda_{jk} \{\le, =, or \ge\} b_i,$$

$$\sum_{k=0}^{r_j} \lambda_{jk} = 1, \qquad j = 1, 2, \ldots, n; \, i = 1, 2, \ldots, m, \qquad (7.5.5)$$

$$\lambda_{jk} \ge 0 \text{ for all } j, k,$$

and only one or two adjacent $\lambda_{jk} > 0$. The problem now becomes one of solving for the λ_{jk}, but, after they are found, we can determine the values for x_j from the relation that for each $j = 1, 2, \ldots, n$,

$$x_j = \sum_{k=0}^{r_j} \lambda_{jk} x_{jk}.$$

(7.5.5) is a standard linear programming problem except for the requirement that for each j, no more than two λ_{jk} (and then they must be adjacent) may be positive. The simplex method may be used with the modification that for a given j, if one λ_{jk} is already in the basis, then at most one other, namely λ_{jk-1} or λ_{jk+1}, may be introduced in the basis.

It must be pointed out that the maximum value of the approximating function so obtained may be merely a local, but not global, maximum, and that the values of the x_j's obtained through the λ_{jk}'s to yield the local maximum for $\bar{z}$ may not even be a feasible solution to the original problem of (7.5.1). On the other hand, the solution of the approximating linear problem will yield a global maximum and also an approximation to the solution of the given nonlinear separable problem, if each of the component functions $f_j(x_j)$ for $j = 1, \ldots, n$ of the objective function $f(x_1, \ldots, x_n)$ is concave and each of the component constraint functions $g_j^i(x_j)$ for $i = 1, \ldots, m$ and $j = 1, \ldots, n$ is either (i) convex (if the inequality is $\leqslant b_i$) or (ii) concave (if the inequality is $\geqslant b_i$) or (iii) linear (if the ith constraint is an equality, that is, if $g_j^i(x_j) = a_{ij} x_j$).

We have discussed the situation for a maximum, but it is easy to extend the discussion to minima with a few modifications. Here, in order for the minimum of the approximating problem to be a global minimum and a valid approximation of the minimum of the given problem, the component functions $f_j(x_j)$ of the objective function all need to be convex, with the same concavity and convexity requirements for the constraints.

The technique can involve many iterations in the simplex algorithm. There are other ways to formulate a continuous nonlinear function in terms of a piecewise linear approximating function. The resulting linear programming problem, however, usually involves many iterations, especially if we use very fine subintervals. An increase in the number of subintervals results in an increase in the number of variables (the λ_{jk}'s) as well as iterations. The reason is that if the initial feasible value of x_j, for a certain j, is far removed from its final value in the optimal solution, it may be necessary to keep moving from one interval to the next one till the final interval is reached, each time necessitating an iteration of the simplex process, although in some cases it is possible to make jumps from one interval to another which is not adjacent to it. In spite of the large number of variables and iterations, the method of approximation is a popular one and several computer programs have been developed for the application of the simplex method to nonlinear separable functions.

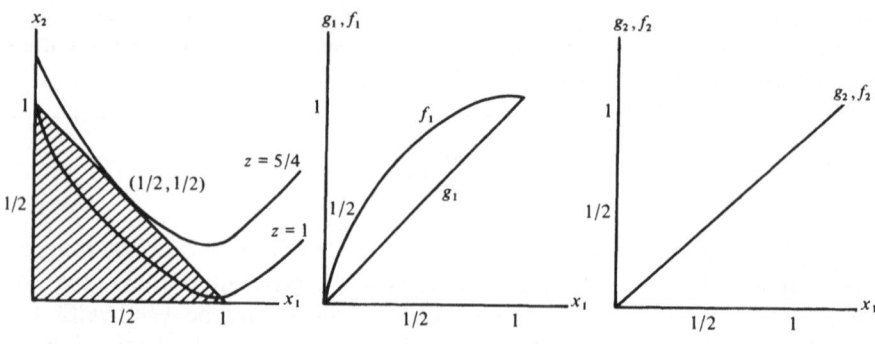

Figure 7.5.2.

Example. Let us consider a very simple problem.
Maximize

$$z = 2x_1 - x_1^2 + x_2$$

(7.5.6)

subject to

$$x_1 + x_2 \leqslant 1, \qquad x_1, x_2 \geqslant 0.$$

The problem may be solved very easily graphically and by realizing that the maximum value of z is obtained at the point where the curve $z = 2x_1 - x_1^2 + x_2$ is tangent to the line $x_1 + x_2 = 1$, or where the slope is the same for both functions, namely -1. Solving for x_1 and x_2 from the equations $x_1 + x_2 = 1$ and

$$\frac{dx_2}{dx_1} = 2x_1 - 2 = -1,$$

we get $x_1 = 1/2$, $x_2 = 1/2$, $z = 5/4$ (see Figure 7.5.2).

The functions $z = f(x_1, x_2)$ and g are separable functions with $f_1(x_1) = 2x_1 - x_1^2$, $f_2(x_2) = x_2$, $g_1(x_1) = x_1$, $g_2(x_2) = x_2$. Moreover, both f_1 and f_2 are concave, and both g_1 and g_2, being linear, are convex. (Notice that there is only one constraint of the type $g^i(x_1, x_2) \leqslant b_i$, so that $i = 1$.) Neither x_j, $j = 1, 2$ can exceed the value 1. For each x_j, we let $r_j = 4$. That is, we have five points $x_{j0} = 0$, $x_{j1} = \frac{1}{4}$, $x_{j2} = \frac{1}{2}$, $x_{j3} = \frac{3}{4}$ and $x_{j4} = 1$, for $j = 1, 2$, and calculate the corresponding values for g_{jk} and f_{jk} for $j = 1, 2$, $k = 0, 1, 2, 3, 4$, as shown in Tables 7.5.1 and 7.5.2.

The approximating linear problem is:
Maximize

$$\bar{z} = \sum_{k=0}^{4} f_{1k}\lambda_{1k} + \sum_{k=0}^{4} f_{2k}\lambda_{2k}$$

$$= 0\lambda_{10} + \tfrac{7}{16}\lambda_{11} + \tfrac{3}{4}\lambda_{12} + \tfrac{15}{16}\lambda_{13} + \lambda_{14} + 0\lambda_{20}$$

$$+ \tfrac{1}{4}\lambda_{21} + \tfrac{1}{2}\lambda_{22} + \tfrac{3}{4}\lambda_{23} + \lambda_{24},$$

Table 7.5.1.

	x_{10}	x_{11}	x_{12}	x_{13}	x_{14}
x_{1k}	0	$\frac{1}{4}$	$\frac{1}{2}$	$\frac{3}{4}$	1
$g_{1k}=x_{1k}$	0	$\frac{1}{4}$	$\frac{1}{2}$	$\frac{3}{4}$	1
$f_{1k}=2x_{1k}-x_{1k}^2$	0	$\frac{7}{16}$	$\frac{3}{4}$	$\frac{15}{16}$	1

subject to

$$\sum_{k=0}^{4} g_{1k}\lambda_{1k} + \sum_{k=0}^{4} g_{2k}\lambda_{2k} \leqslant 1 \quad \text{and} \quad \sum_{k=0}^{4} \lambda_{jk}=1, \quad j=1,2.$$

Or, after the introduction of a slack variable x_3, the constraints are

$$0\lambda_{10}+\tfrac{1}{4}\lambda_{11}+\tfrac{1}{2}\lambda_{12}+\tfrac{3}{4}\lambda_{13}+\lambda_{14}+0\lambda_{20}+\tfrac{1}{4}\lambda_{21}$$

$$+\tfrac{1}{2}\lambda_{22}+\tfrac{3}{4}\lambda_{23}+\lambda_{24}+x_3=1, \tag{7.5.7}$$

$$\sum_{k=0}^{4} \lambda_{1k}=1, \quad \sum_{k=0}^{4} \lambda_{2k}=1, \quad \lambda_{jk}\geqslant 0, \quad \text{for } j=1,2; k=0,1,2,3,4,$$

and for each j at most two λ_{jk} (and then they must be adjacent) may be positive at any one iteration.

The iterations are shown in Table 7.5.3. Since $g_{10}=g_{20}=0$, the initial basic variables are λ_{10}, λ_{20}, and x_3, as shown in Iteration 1. In the $\bar{z}$-row, the coefficients of λ_{14} and λ_{24} are both -1. We choose λ_{14} as the entering basic variable, and λ_{10} as the leaving, as shown in Iteration 2. Notice that although x_3 appears to be tied with λ_{10} as the leaving variable, λ_{10} has to be the one to leave, because it is not adjacent to λ_{14}.

Although the coefficient of λ_{24} in the next $\bar{z}$-row is -1, we cannot enter λ_{24}, because, if we were to do so, x_3 would have to leave the basis (the ratio $0/1$ being smallest), and λ_{24} is not adjacent to λ_{20}, which is already in the basis. Similarly we cannot enter λ_{23} or λ_{22}, but λ_{21} is adjacent to λ_{20} and can be entered, replacing x_3 in the basis. This is shown in Iteration 3.

At Iteration 4, if λ_{12} were entered, λ_{20} would have to leave (the ratio $1/2$ being smaller than $1/1$) but we cannot allow both λ_{12} and λ_{14} in the basis. So λ_{13} is entered, replacing λ_{14}.

Table 7.5.2.

	x_{20}	x_{21}	x_{22}	x_{23}	x_{24}
x_{2k}	0	$\frac{1}{4}$	$\frac{1}{2}$	$\frac{3}{4}$	1
$g_{2k}=x_{2k}$	0	$\frac{1}{4}$	$\frac{1}{2}$	$\frac{3}{4}$	1
$f_{2k}=x_{2k}$	0	$\frac{1}{4}$	$\frac{1}{2}$	$\frac{3}{4}$	1

Table 7.5.3

| Iteration | Basis | Current coefficient | | | | | | | | | | | Current Value | Ratio |
		λ_{10}	λ_{11}	λ_{12}	λ_{13}	λ_{14}	λ_{20}	λ_{21}	λ_{22}	λ_{23}	λ_{24}	x_3		
1	$\bar{z}$	0	$-\frac{7}{16}$	$-\frac{3}{4}$	$-\frac{15}{16}$	-1	0	$-\frac{1}{4}$	$-\frac{1}{2}$	$-\frac{3}{4}$	-1	0	0	
	x_3	0	$\frac{1}{4}$	$\frac{1}{2}$	$\frac{3}{4}$	1	0	$\frac{1}{4}$	$\frac{1}{2}$	$\frac{3}{4}$	1	1	1	1
	λ_{10}	1	1	1	1	1	0	0	0	0	0	0	1	1
	λ_{20}	0	0	0	0	0	1	1	1	1	1	0	0	
2	$\bar{z}$	1	$\frac{9}{16}$	$\frac{1}{4}$	$\frac{1}{16}$	0	0	$-\frac{1}{4}$	$-\frac{1}{2}$	$-\frac{3}{4}$	-1	0	1	
	x_3	-1	$-\frac{3}{4}$	$-\frac{1}{2}$	$-\frac{1}{4}$	0	0	$\frac{1}{4}$	$\frac{1}{2}$	$\frac{3}{4}$	1	1	0	0
	λ_{14}	1	1	1	1	1	0	0	0	0	0	0	1	
	λ_{20}	0	0	0	0	0	1	1	1	1	1	0	1	1
3	$\bar{z}$	0	$-\frac{3}{16}$	$-\frac{1}{4}$	$-\frac{3}{16}$	0	0	0	0	0	0	1	1	
	λ_{21}	-4	-3	-2	-1	0	0	1	2	3	4	4	0	
	λ_{14}	1	1	1	1	1	0	0	0	0	0	0	1	1
	λ_{20}	4	3	2	1	0	1	0	-1	-2	-3	-4	1	1
4	$\bar{z}$	$\frac{3}{16}$	0	$-\frac{1}{16}$	0	$\frac{3}{16}$	0	0	0	0	0	1	$\frac{19}{16}$	
	λ_{21}	-3	-2	-1	0	1	0	1	2	3	4	4	1	
	λ_{13}	1	1	1	1	1	0	0	0	0	0	0	1	1
	λ_{20}	3	2	1	0	-1	1	0	-1	-2	-3	-4	0	0
5	$\bar{z}$	$\frac{6}{16}$	$\frac{2}{16}$	0	0	$\frac{2}{16}$	$\frac{1}{16}$	0	$-\frac{1}{16}$	$-\frac{3}{16}$	$-\frac{3}{16}$	$\frac{13}{16}$	$\frac{19}{16}$	
	λ_{21}	0	0	0	0	0	1	1	1	1	1	0	1	1
	λ_{13}	-2	-1	0	1	2	-1	0	1	2	3	4	1	1
	λ_{12}	3	2	1	0	-1	1	0	-1	-2	-3	-4	0	
6	$\bar{z}$	$\frac{1}{4}$	$\frac{1}{16}$	0	$\frac{1}{16}$	$\frac{1}{4}$	0	0	0	0	0	1	$\frac{5}{4}$	
	λ_{21}	2	1	0	-1	-2	2	1	0	-1	-2	-4	0	
	λ_{22}	-2	-1	0	1	2	-1	0	1	2	3	4	1	
	λ_{12}	1	1	1	1	1	0	0	0	0	0	0	1	

At Iteration 5, λ_{12} is the one to enter. At Iteration 6, λ_{24} or λ_{23} cannot be entered because neither is adjacent to λ_{21}, but λ_{22} may, replacing λ_{13}.[10] The process terminates with $\lambda_{12}=1$, $\lambda_{22}=1$, $\bar{z}=5/4$, all other λ_{jk}'s $=0$.

Since $x_1=\sum_{k=0}^{4}\lambda_{1k}x_{1k}$ and $x_2=\sum_{k=0}^{4}\lambda_{2k}x_{2k}$, we get $x_1=\lambda_{12}x_{12}=\frac{1}{2}$ and $x_2=\lambda_{22}x_{22}=\frac{1}{2}$, which turns out to be the same as the solution obtained graphically and is the solution to (7.5.6). $\square$

---

[10]If the reader chooses λ_{21}, instead of λ_{13}, as the leaving variable, he will arrive at the same solution after one more iteration.

Sometimes a function which is originally not separable may be converted into one which is. The objective function of a quadratic programming problem, for instance, is not separable because it contains the products $x_i x_j$. An $x_i x_j$ term may be eliminated quite easily by introducing two variables, say, y_i and y_j, with

$$y_i = \tfrac{1}{2}(x_i + x_j) \quad \text{and} \quad y_j = \tfrac{1}{2}(x_i - x_j). \tag{7.5.8}$$

Then $x_i x_j = y_i^2 - y_j^2$. For example, suppose we have an objective function

$$f(x_1, x_2) = x_1^2 + x_1 x_2 + 2x_2^2.$$

It becomes

$$\varphi(x_1, x_2, y_1, y_2) = x_1^2 + y_1^2 - y_2^2 + 2x_2^2$$

after the substitution of (7.5.8). The resulting separable problem would have the two additional variables y_1 and y_2 and the two additional constraints of (7.5.8). Moreover, y_2, being the difference of x_1 and x_2, will be unrestricted in sign. But the simplex algorithm may be applied as previously discussed. In fact, if we use the λ-method of approximation presented in this section, the λ's will still have to be nonnegative and restricted in that only two adjacent ones may be positive.

If it is known from the nature of the problem that the variables are strictly positive, then products of the form $x_i x_j x_k$ involving three or more variables may be separated by the introduction of the variable

$$y = x_i x_j x_k.$$

Replace the product $x_i x_j x_k$ by y and add the separable constraint

$$\log y = \log x_i + \log x_j + \log x_k.$$

We have mentioned just one or two simple examples of how separability may be achieved through the transformation of variables. There are other ways to handle less simple functions. The size of the approximating problem probably becomes very large with the introduction of many additional variables and constraints, but can be handled by computer codes, and the technique of approximating a nonlinear problem by piecewise linear functions is one frequently used.

We have looked at quadratic programming and separable functions in this chapter not only because of their wide applicability but also because they further illustrate how the simplex method of linear programming is used in nonlinear problems.

7.6 Some economic applications of the Kuhn–Tucker conditions

Classical methods of optimization under constraint as discussed in Chapter 3 cannot be used if the number of constraints exceeds the number of variables or if the constraints are in the form of inequalities. In many problems in economics, physical or practical considerations are such that the Kuhn–Tucker conditions can be a useful aid in arriving at solutions. Let us look at the following illustrations.

Example 1. Peak-load pricing.[11] With rising fuel costs, many utilities, such as the Detroit Edison, may find it worthwhile to employ a pricing system in which customers are charged a higher rate during peak periods of use and a lower rate during off-peak periods. Disregarding the energy conservation arguments, the company is interested in knowing whether such a pricing system maximizes profit.

Let: $q_1, q_2, \ldots, q_{24}$ be the quantity (of electricity, say) sold or demanded during each of the 24 hours of the day; $p_1, p_2, \ldots, p_{24}$ be the corresponding prices per unit; y be the hourly output capacity; $C(q_1, q_2, \ldots, q_{24})$ be the daily total operating cost; and $g(y)$ be the daily cost of capital (capacity).

The problem, then, is to maximize daily profit $z = f(q_i, \ldots, q_{24})$, where

$$z = \sum_{i=1}^{24} p_i q_i - C(q_1, \ldots, q_{24}) - g(y),$$

subject to

$$0 < q_i \leqslant y, \qquad i = 1, 2, \ldots, 24.$$

Notice that we assume that each $q_i > 0$, that is, some quantity is sold or demanded each hour of the day, and that $y > 0$. We also assume that each p_i charged is independent of the output sold, *i.e.*, $\partial p_i / \partial q_i = 0$.

It follows from the Kuhn–Tucker (K–T) conditions in (7.2.2) and (7.2.3) that there exist n nonnegative λ's $\lambda_1, \lambda_2, \ldots, \lambda_{24}$ such that

$$\frac{\partial z}{\partial q_i} = p_i - \frac{\partial C}{\partial q_i} - \lambda_i = 0 \quad \text{for } i = 1, 2, \ldots, 24 \tag{7.6.1}$$

and

$$\frac{\partial z}{\partial y} = -\frac{dg(y)}{dy} + \sum_{i=1}^{24} \lambda_i = 0. \tag{7.6.2}$$

Moreover, the K–T condition given by (7.2.4) requires that

$$\frac{\partial z}{\partial \lambda_i} = y - q_i \geqslant 0 \tag{7.6.3}$$

and

$$\lambda_i(y - q_i) = 0 \quad \text{for } i = 1, \ldots, 24.$$

If it is an off-peak period, we must have $y > q_i$; hence, from (7.6.3), $\lambda_i = 0$. Putting this into (7.6.1), we reach the conclusion that, during off-peak periods,

$$p_i = \frac{\partial C}{\partial q_i},$$

[11]Adapted from William Baumol (1972), pp. 167–170.

that is, the optimal price should be equal to the marginal operating cost.

On the other hand, during a peak period, say t, we may have $y = q_t$, so that λ_t may be > 0. And, since (7.6.1) tells us that

$$p_t = \frac{\partial C}{\partial q_t} + \lambda_t,$$

the price during any peak period t should optimally exceed the marginal operating cost by a supplementary amount equal to λ_t, and, from (7.6.2),

$$\frac{dg}{dy} = \sum_{i=1}^{24} \lambda_i,$$

that is, the sum of these supplementary amounts for all peak periods should be equal to the marginal capacity cost.

Hence, the above analysis justifies charging a higher rate during peak periods, from the point of view of profit maximization.

Example 2 Allocating joint costs.[12] It often happens that a raw material is used simultaneously in the manufacture of several products. For instance, soy beans are cooked and the juice coagulated and made into bean curd, which is a major food item in the Far East. The pulp is used either as hog feed or fertilizer. How does one allocate the joint costs of these products? A traditional way is to determine the marginal revenue of each product in a profit-maximizing plan and call the marginal revenues "costs". This approach, however, does not allocate the original cost of the raw material to the various products.

An alternative method is to use the K–T technique. Suppose soy beans cost $10 per 100-lb. bag. Let:

$B =$ number of bags of soy beans purchased;
$q_1 =$ number of units of bean curd obtained from the B bags of beans;
$p_1 =$ unit price of bean curd;
$q_2 =$ number of units of fertilizer obtained;
$p_2 =$ unit price of fertilizer.

We wish to maximize the profit $z = f(q_1, q_2, B)$, where

$$z = p_1 q_1 + p_2 q_2 - 10B,$$

subject to $0 < q_i \leqslant B$, $i = 1, 2$.
Notice again that we assume that we buy some beans and get bean curd and fertilizer as products. Suppose the demand functions are

$$p_1 = 40 - 2q_1 \quad \text{and} \quad p_2 = 20 - q_2.$$

Then

$$z = 40q_1 - 2q_1^2 + 20q_2 - q_2^2 - 10B,$$

---

[12]Adapted from Weil, Roman L., 1968. Allocating joint costs, *American Economic Review*, **58**, no. 5, pp. 1342–45.

and the K–T conditions tell us that for profit maximization we must have $\lambda_1 \geqslant 0$ and $\lambda_2 \geqslant 0$ such that

$$\frac{\partial z}{\partial q_1} = 40 - 4q_1 - \lambda_1 = 0, \tag{7.6.4}$$

$$\frac{\partial z}{\partial q_2} = 20 - 2q_2 - \lambda_2 = 0, \tag{7.6.5}$$

$$\frac{\partial z}{\partial B} = -10 + \lambda_1 + \lambda_2 = 0, \tag{7.6.6}$$

$$B - q_i \geqslant 0 \quad \text{and} \quad \lambda_i(B - q_i) = 0 \quad \text{for } i = 1, 2. \tag{7.6.7}$$

From the above we conclude that neither $\lambda_i = 0$. For suppose $\lambda_1 = 0$. Then we must have $\lambda_2 = 10$, $q_1 = 10$, $q_2 = 5$. But, from (7.6.7), since $\lambda_2 \neq 0$, $B = q_2$ $= 5$, contradicting the requirement $B - q_i \geqslant 0$. So $\lambda_1 > 0$, $\lambda_2 > 0$, and $B = q_1$ $= q_2$. This enables us to solve the three equations given by (7.6.4)–(7.6.6) to get

$$B = q_1 = q_2 = 8\tfrac{1}{3}, \qquad \lambda_1 = 6\tfrac{2}{3}, \quad \text{and} \quad \lambda_2 = 3\tfrac{1}{3}.$$

That is, the most profitable action would be to buy $8\tfrac{1}{3}$ bags of beans from which can be produced an equal amount (in units) of bean curd and fertilizer. And one would allocate $\lambda_1 = \$6\tfrac{2}{3}$ to the curd-part cost of the beans and $\lambda_2 = \$3\tfrac{1}{3}$ to the fertilizer part, with the sum of these two costs equal to \$10, which is the cost of the beans.

This technique can be used to solve more complicated joint cost problems. Suppose one raw material R is used in the production of three products A, B, and C. To produce B and C requires a special processing of R. Let:

x_1, x_2, and x_3 be the quantities of A, B, and C produced;
$f_1(x_1)$, $f_2(x_2)$, and $f_3(x_3)$ be the respective price functions;
$R = $ amount of raw material bought;
$y = $ amount of R processed for the production of B and C.

Suppose further that the unit cost of R is \$10 and the (joint) cost of processing each unit of y is \$4. Our problem is to maximize

$$z = \sum_{i=1}^{3} x_i f_i(x_i) - 10R - 4y,$$

subject to

$$0 < x_1 \leqslant R, \qquad 0 < x_2 \leqslant y, \qquad 0 < x_3 \leqslant y, \qquad 0 < y \leqslant R.$$

Our associated Lagrangian function (see Section 7.3) is

$$L(\mathbf{x}, R, y, \boldsymbol{\lambda}) = \sum_{i=1}^{3} x_i f_i(x_i) - 10R - 4y + \lambda_1(R - x_1)$$

$$+ \lambda_2(y - x_2) + \lambda_3(y - x_3) + \lambda_4(R - y).$$

Since neither R nor $y=0$, we have, from the K–T condition in (7.2.4),

$$\frac{\partial L}{\partial R} = -10+\lambda_1+\lambda_4=0 \tag{7.6.8}$$

and

$$\frac{\partial L}{\partial y} = -4+\lambda_2+\lambda_3-\lambda_4=0. \tag{7.6.9}$$

Without the explicit demand functions $f_i(x_i)$ we cannot, of course, determine the values of the λ_i. Nevertheless, we see from (7.6.8) that we allocate the \$10 cost of R between λ_1 for A and λ_4 for B and C jointly. The \$4 cost of further processing plus the λ_4 amount already allocated for B and C jointly are then divided between λ_2 for B and λ_3 for C, as seen in (7.6.9). ☐

This example, by the way, reinforces the concepts developed in Sections 6.3 and 7.3 that these λ's are opportunity or implicit costs, or imputed or shadow prices.

As these two examples show, the Kuhn–Tucker conditions, by virtue of the requirements

$$x_j^o\left[\frac{\partial f(\mathbf{x}^o)}{\partial x_j} - \sum \lambda_i^o \frac{\partial g^i(\mathbf{x}^o)}{\partial x_j} \right]=0 \quad \text{and} \quad \lambda_i^o\left[b_i - g^i(\mathbf{x}^o) \right]=0,$$

enable one to solve equations (instead of inequalities) for the x_j's or λ_i's when it is known, through physical considerations, that either x_j or λ_i must be nonzero. The technique therefore has wider applicability than the classical approach of Lagrange multipliers, which can only handle situations in which the number of constraints must not exceed the number of variables and must be expressed as equations. The Kuhn–Tucker conditions are used extensively to develop economic theory, such as the theory of investment and production.

Example 3. A pay-off period theory of investment.[13] It is common practice in the investment decision process to limit investment outlays to those projects that will return the investment within a specified pay-off period. Suppose a firm is considering n alternative investment projects, and that the ith project yields a discounted expected profit flow $P_i(K_i)$, where K_i is the investment expenditures on the ith project. Suppose further that the firm has a fixed amount of capital C to distribute among the n projects. The problem is to maximize

$$P= \sum_{i=1}^{n} P_i(K_i),$$

[13]See Smith, Vernon L., 1966, *Investment and production*, pp. 219–222. Harvard University Press, Cambridge, Mass.

subject to $\Sigma_i K_i \leqslant C$, $K_i \geqslant 0$, for $i = 1, \ldots, n$. Letting

$$L(\mathbf{K}, \lambda, C) = \sum_{i=1}^{n} P_i(K_i) + \lambda(C - \Sigma K_i),$$

the K–T conditions tell us that

$$\frac{\partial P_i}{\partial K_i} - \lambda \leqslant 0 \qquad\qquad (7.6.10)$$

and that

$$K_i \left(\frac{\partial P_i}{\partial K_i} - \lambda \right) = 0.$$

We may interpret $\partial L / \partial C = \lambda$ as the marginal profitability of capital C, and say that the marginal pay-off period for project i is the period of time required for an increment of net profit from an increment of investment in that project to pay off the additional investment. Then we may define

$$\phi_i = \frac{1}{\dfrac{\partial P_i}{\partial K_i}}$$

to be the marginal pay-off period on the ith project. Similarly, the equilibrium pay-off period $\bar{\phi}$ may be defined from the marginal profitability of capital to the firm as

$$\bar{\phi} = \frac{1}{\dfrac{\partial L}{\partial C}} = \frac{1}{\lambda}.$$

This leads to the following restatement of (7.6.10):

if $K_i > 0$,   then $\phi_i = \bar{\phi}$,   and

if $\phi_i > \bar{\phi}$,   then $K_i = 0$,   for $i = 1, \ldots, n$.

That is, if the pay-off period exceeds a specified minimum cut-off period $\bar{\phi}$, that project should not be undertaken. Also, according to the K–T conditions, the optimal plan would require $\phi_i \geqslant \bar{\phi}$. Thus if the marginal pay-off period ϕ_i for some project i is below $\bar{\phi}$ ($\phi_i < \bar{\phi}$), that ith project should be expanded till the pay-off on the last unit of investment is equal to $\bar{\phi}$.

Example 4. Cost minimization under capital rationing.[14] Suppose that two inputs, one current (x_1) and one capital (x_2), are involved in a productive process and that the firm has a "ration" of money capital available which cannot exceed a fixed amount, say $\bar{K}$ units. Let:

$w_1 =$ price of each unit of current input;
$w_2 =$ price of each unit of capital input;
$y =$ output.

[14]See Vernon Smith, *op. cit.*, pp. 192–196.

Then we wish to minimize the total cost C, where

$$C = w_1 x_1 + w_2 x_2,$$

subject to $y = f(x_1, x_2)$ and $w_2 x_2 \leqslant \overline{K}$, $x_1 > 0$, $x_2 > 0$. To minimize C is the same as to maximize $-C$, and we can write the associated Lagrangian as

$$L = -w_1 x_1 - w_2 x_2 + \lambda_1 (f - y) + \lambda_2 (\overline{K} - w_2 x_2),$$

where $\lambda_2 \geqslant 0$ and λ_1 is unrestricted in sign. The economic interpretation of the λ's is: λ_1 is the marginal cost of output, and λ_2 is the marginal cost of money capital. By the K–T conditions, we get

$$\frac{\partial L}{\partial x_1} = -w_1 + \lambda_1 \frac{\partial f}{\partial x_1} = 0, \qquad (7.6.11)$$

$$\frac{\partial L}{\partial x_2} = -w_2 + \lambda_1 \frac{\partial f}{\partial x_2} - \lambda_2 w_2 = 0, \qquad (7.6.12)$$

$$\frac{\partial L}{\partial \lambda_1} = f - y \geqslant 0, \lambda_1 (f - y) = 0, \qquad (7.6.13)$$

$$\frac{\partial L}{\partial \lambda_2} = \overline{K} - w_2 x_2 \geqslant 0, \lambda_2 (\overline{K} - w_2 x_2) = 0. \qquad (7.6.14)$$

If $\overline{K} - w_2 x_2 > 0$, then $\lambda_2 = 0$, i.e., the marginal cost of money capital is zero, meaning that the firm has more money capital than it can use. Consequently, the output $y = f(x_1, x_2)$ can be expanded without any capital constraint. And if $\lambda_2 > 0$, then $\overline{K} - w_2 x_2 = 0$, and money capital is a constraint in the maximizing process. From the four equations given by (7.6.11)–(7.6.14), we can solve for the optimal values

$$x_1^o = h_1(x_1, x_2, K, y),$$
$$x_2^o = h_2(x_1, x_2, K, y),$$
$$\lambda_1 = h_3(x_1, x_2, K, y),$$
$$\lambda_2 = h_4(x_1, x_2, K, y).$$

Exercises

7.1 For the problem: Minimize $z = (x_1 - 4)^2 + (x_2 - 3)^2$, subject to the constraints given in (7.1.1); verify that $(34/13, 17/13)$ is an optimal solution by applying the Kuhn–Tucker conditions and finding the values of λ_i.

7.2 Maximize $z = 2x_1 - x_1^2 + x_2$, subject to $x_1 + x_2 \leqslant 3$, $3x_1 - 2x_2 \leqslant 6$, $x_1, x_2 \geqslant 0$.

(1) Solve the problem graphically.
(2) Verify your answer by applying the Kuhn–Tucker conditions and finding the values of λ_i.
(3) Solve the problem by using the quadratic programming technique discussed in Section 7.4.
(4) Solve the problem by using the approximation method of Section 7.5.
(5) Formulate the dual to this problem and verify that $\max z = \min w$.

7.3 (1) Maximize $z = x_1 + x_2 - x_1^2 + x_1 x_2 - \frac{1}{2} x_2^2$, subject to $x_1 + x_2 \leqslant 3$, $3x_1 + 2x_2 \geqslant 6$, $x_1, x_2 \geqslant 0$.

   (2) Verify your answer by checking the signs of the λ's.

   (3) Formulate the dual and verify that $\max z = \min w$.

7.4 Verify Table 7.5.3. Choose λ_{21}, instead of λ_{13}, as the leaving variable at Iteration 6.

7.5 The management of a firm wishes to maximize sales revenue $R = R(q)$, subject to $\pi = R(q) - C(q) \geqslant \pi_0$, $q \geqslant 0$, $\pi_0 > 0$, where q, π, and C stand for output, profit and cost respectively, and π_0 is a given profit level. Use the Kuhn–Tucker conditions to derive the sales maximizing output rule.

7.6 Let the firm of the problem above have revenue and cost functions

$$R(q) = 20q - q^2 \quad \text{and} \quad C(q) = q^2 - 4q + 14,$$

and let minimum profit be $\pi_0 = 50$. Find the sales maximizing output level.

8 Optimal Control

8.1 The control problem and some terminology

The optimization techniques we have considered thus far are *static* in the sense that time does not enter into consideration. By contrast, *dynamic* optimization involves optimization over an interval of time. It is obvious that in many decision-making situations, time is a factor which cannot be ignored.

As an example,[1] consider a firm which wishes to maximize its total profits over a certain period of time, say from t_0 to t_f. At any given moment of time t, $t_0 \leqslant t \leqslant t_f$, the firm will have a certain stock of capital and other factors of production. Let us ignore the other factors of production for the time being and think only of capital, and denote it by $k(t)$, since the amount of capital is a function of time t. We shall call $k(t)$ a *state* variable, as it represents the state (as far as capital stock is concerned, thought of as the only factor of production) in which the firm finds itself at time t. At that time t, the firm will make a decision on, say, the rate of production. Call this decision $u(t)$. From the available amount of capital k together with the decision u at the specified time t, the firm makes a certain profit per unit of time. Since this profit earned at time t is the result of capital k and decision u, we denote it by $I[k(t), u(t), t]$. If we regard time as continuous, it then follows that the total profit for that firm earned from time t_0 to time t_f is given by

$$J = \int_{t_0}^{t_f} I[k(t), u(t), t] \, dt. \tag{8.1.1}$$

The decision policy varies over time, and, to show the time path that it takes during the period from t_0 to t_f some writers emphasize this by writing $\vec{u}(t)$. The firm chooses the decision policy $\vec{u}(t)$, subject to certain con-

[1] This example is based on Robert Dorfman (1969).

straints, and that determines the rate of change of the capital stock. That is to say, the time rate of change of the capital stock at any instant, which is dk/dt, is a function of the present state or amount of capital available $k(t)$, the decision $u(t)$ taken, and the specific time t. In symbols, we have

$$\dot{k} = \frac{dk}{dt} = f(k,u,t).^2 \tag{8.1.2}$$

From this we see that the decision variable $u(t)$ influences or controls the rate at which the capital stock changes and thereby the amount of capital stock available at later instants of time. The total profit, then, is really a function of the decision policy $u(t)$, and the objective of the firm is to maximize total profit with respect to $u(t)$, that is,

$$\max_{u(t)} J[u(t)] = \int_{t_0}^{t_f} I(k,u,t)\,dt.$$

We shall refer to $u(t)$ as the *control* variable.

Usually a firm has, besides capital stock, other factors of production such as labor, *etc.*, and will be making decisions over not just the rate of production but such other matters as price of output (if the firm is a monopoly), *etc.* We then represent the state variables and control variables by vectors, so that in this chapter $\mathbf{x}(t)$ will stand for a column vector of state variables (such as capital stock, labor, *etc.*) and $\mathbf{u}(t)$ a column vector of control variables (such as rate of production, price of output, *etc.*).

$$\mathbf{x}(t) = \begin{bmatrix} x_1(t) \\ x_2(t) \\ \vdots \\ x_n(t) \end{bmatrix}, \quad \mathbf{u}(t) = \begin{bmatrix} u_1(t) \\ u_2(t) \\ \vdots \\ u_r(t) \end{bmatrix}, \quad \dot{\mathbf{x}} = \frac{d\mathbf{x}}{dt} = f(\mathbf{x}, \mathbf{u}, t) = \begin{bmatrix} dx_1/dt \\ dx_2/dt \\ \vdots \\ dx_n/dt \end{bmatrix}.$$

The total profit J is called the *objective* or *criterion functional* and is a real-valued scalar. A functional may be thought of as the function of a function. It differs from a composite function in the following manner. Suppose

$$z = f(y) \quad \text{and} \quad y = g(x)$$

are functions; then

$$z = f(g(x)) = h(x)$$

is a composite function. For a given value x in an interval $x_0 \leq x \leq x_1$, there is one unique value $y = g(x)$, as specified by the relation g. On the other hand, if there is a class of functions $\{y\}$ on the interval $[x_0, x_1]$, (for

---

[2]The symbol $\dot{x}$ is commonly used in the literature to stand for the time rate of change of a variable x; that is, $\dot{x} = dx/dt$. For higher derivatives, the notation is $\ddot{x} = d^2x/dt^2$, $\dddot{x} = d^3x/dt^3$, *etc.* Also note that in (8.1.2) we have written $f(k,u,t)$, instead of $f[k(t),u(t),t]$ as we should, to avoid a cluttered look. This practice will be followed whenever we believe there is no confusion in realizing that k and u are functions of t.

instance, let the class be C^1, the set of all continuously differentiable functions on $[x_0, x_1]$), y is any member in that class, and z is a function of y, then we say z is a functional. We indicate this by writing

$$z = f(y), \qquad y = g(x), \quad \text{and } g(x) \in C^1.$$

In our example, the control variable $\mathbf{u}(t)$ will be restricted to belong to some set $\mathscr{U}$ of functions on $[t_0, t_f]$. The functional J may be a function of any $\mathbf{u} \in \mathscr{U}$, and we seek that $\mathbf{u}$ which maximizes J.

This leads us to a formal statement of the general control problem, which is to maximize a functional $J[\mathbf{u}(t)]$, subject to certain conditions:

$$\max_{\mathbf{u}(t)} J[\mathbf{u}(t)] = \int_{t_0}^{t_f} I[\mathbf{x}(t), \mathbf{u}(t), t] \, dt + F(\mathbf{x}_f, t_f)$$

subject to

$$\dot{\mathbf{x}} = \frac{d\mathbf{x}}{dt} = f[\mathbf{x}(t), \mathbf{u}(t), t], \tag{8.1.3}$$

where:

t_0 and $\mathbf{x}_0 = \mathbf{x}(t_0)$ are given;
$[\mathbf{x}(t), t] \in T$ at $t = t_f, \mathbf{x}_f = x(\mathbf{t}_f)$;
$\mathbf{u}(t) \in \mathscr{U} = \{\mathbf{u}(t) | \mathbf{u}: [t_0, t_f] \to \Omega, \text{ piecewise continuous}\}$:

T is some given terminal surface; Ω is a nonempty subset of Euclidean r space $\mathbf{R}^r$; and the functions I and F are assumed given and continuously differentiable.

Notice that the functional J is a sum of two terms. This is the most general form; either term may be zero, as will be mentioned again later.

In (8.1.3), $\mathbf{x}$ is an $n \times 1$ vector and is called a *state* vector[3] of the system, and the n real numbers $x_1(t), x_2(t), \ldots, x_n(t)$ are called *state* variables or *phase* coordinates of the controlled object or process. Thus the n-dimensional vector $\mathbf{x}$ may be thought of as a point in Euclidean n-space at time t and is also referred to as a *phase* point. We shall assume each $x_i(t)$, $i = 1, \ldots, n$ to be a continuously differentiable function of t. The *control* vector $\mathbf{u}(t)$ is an r-dimensional vector which may be thought of as a point in Euclidean r-space $\mathbf{R}^r$, at time t. The r control variables $u_j(t)$, $j = 1, 2, \ldots, r$, will be subject to certain constraints.[4] Let Ω be a given nonempty set in $\mathbf{R}^r$. We define a control vector $\mathbf{u}$ to be *admissible* if (i) it is piecewise continuous, that is, if it is continuous over the interval $[t_0, t_f]$ except for a finite number of jumps, and (ii) its value at any time in the interval $[t_0, t_f]$ belongs to Ω, that is,

$$\mathbf{u}(t) \in \Omega \subset \mathbf{R}^r \quad \text{for } t_0 \leqslant t \leqslant t_f.$$

---

[3]Sometimes we shall refer to it as a state variable when there can be no confusion.
[4]For example, if u is the price of an output, the firm may decide that during the time interval $[t_0, t_f]$ the price should lie within a certain range.

We shall denote the set of all such admissible control vectors by $\mathscr{U}$. Thus $\mathbf{u} \in \mathscr{U}$ means u satisfies the above two conditions.

In order to emphasize time as the argument of the functions, we refer to them as the *control trajectory* (or time path of the control vector) or the *state trajectory* (or time path of the state vector) respectively.

The state trajectory is characterized by certain *boundary* conditions such as

$$\mathbf{x}_0 = \mathbf{x}(t_0) \quad \text{and} \quad \mathbf{x}_f = \mathbf{x}(t_f)$$

or $g(\mathbf{x}_0, \mathbf{x}_f, t_0, t_f) = 0$ for some function g (where t_0 and t_f represent initial time and terminal time respectively) and by the n differential *equations of motion*

$$\dot{\mathbf{x}} = f(\mathbf{x}, \mathbf{u}, t), \tag{8.1.4}$$

or, if we write out the components of the vectors,

$$\dot{x}_i = \frac{dx_i}{dt} = f_i\big[\, x_1(t), x_2(t), \dots, x_n(t), u_1(t), \dots, u_r(t), t \,\big], \qquad i = 1, 2, \dots, n.$$

If t does not enter explicitly as an argument of f, that is, if $\dot{\mathbf{x}}$ does not depend explicitly on time so that, instead of (8.1.4), we have

$$\dot{\mathbf{x}} = f\big[\mathbf{x}(t), \mathbf{u}(t)\big],$$

we say the system is *autonomous*.

While the initial time t_0 and initial state $\mathbf{x}_0$ are usually given, the terminal time t_f and terminal state $\mathbf{x}_f$ are usually to be determined through a specified *terminal surface* T which is a subset of the Euclidean $n+1$-space $\mathbf{R}^{n+1}$. For instance, if there is only one state variable (so that the vector $\mathbf{x}$ is the one-dimensional scalar function x), T would represent a curve, as shown in Figure 8.1.1. T is a specified relation between $x(t)$ and t at $t = t_f$. When this is the case, the optimal trajectory has to necessarily satisfy these additional *transversality* conditions for the terminal boundary requirements.

A *feasible* state vector $\mathbf{x}$ is one which satisfies the boundary (initial and terminal) conditions and is obtained from the equations of motion through an admissible control $\mathbf{u}$.

In Figure 8.1.1, several feasible state trajectories are drawn for the case where there is only one state variable x. If x_a is the optimal trajectory, the terminal point would be A and the terminal time $t_{f,a}$. But if x_b is the optimal trajectory, the terminal point would be B and the terminal time $t_{f,b}$. Notice that both terminal points lie on the terminal curve T.

In (8.1.3), the integrand $I(\mathbf{x}, \mathbf{u}, t)$ is called the *intermediate function*. The second term, $F(\mathbf{x}_f, t_f)$, is called the *final function* and shows the dependence of the objective functional J on the terminal time t_f. The problem as stated in (8.1.3) is frequently referred to as the *Problem of Bolza*. If the final function is zero so that

$$J = \int_{t_0}^{t_f} I(\mathbf{x}, \mathbf{u}, t) \, dt,$$

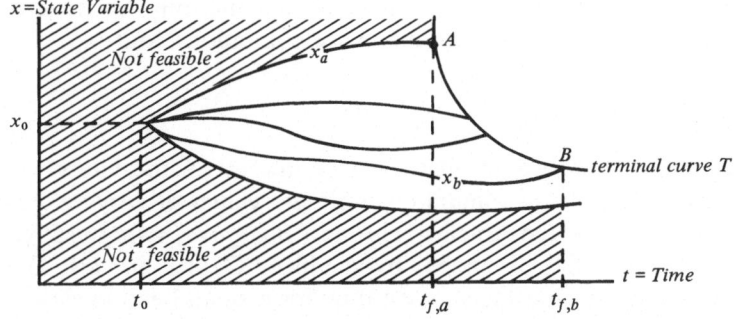

Figure 8.1.1.

it is called the *Problem of Lagrange*, while if the intermediate function is zero so that

$$J = F(\mathbf{x}_f, t_f),$$

it is called the *Problem of Mayer*. It can be easily shown that, by suitable substitution, one type of problem can be rewritten so as to be of another type.

We have stated the problem as one of maximization. This is solely for the sake of convenience. As we have said previously, to minimize a function or a functional is equivalent to maximizing the negative of that function or functional. In what follows, we shall be referring mostly to a maximization problem, but the reader may bear in mind that the same discussion, with an appropriate change of sign, is applicable to a minimization problem.

The control problem as stated in (8.1.3) is usually approached by one of two techniques, one called the calculus of variations and the other dynamic programming. These two approaches both use the concept of variations and differ mainly in emphasis, as will be more apparent later. The modern concept of variational calculus is much more general than what is called the classical calculus of variations, and it is often referred to as the *maximum principle* or *Pontryagin principle*. We shall, however, first consider the classical calculus of variations, because the technique well illustrates the variational nature of the approach and is relatively easy to use in some special cases.

8.2 The classical calculus of variations

The problem that is treated in the framework of the classical calculus of variations is one of Lagrange with the control vector equal to the time derivative of the state vector. That is, the problem is:

$$\max_{\dot{\mathbf{x}}(t)} J[\dot{\mathbf{x}}(t)] = \int_{t_0}^{t_f} I[\mathbf{x}(t), \dot{\mathbf{x}}(t), t] dt \qquad (8.2.1)$$

where t_0 and $\mathbf{x}_0$ are given, and $I(\mathbf{x}, \dot{\mathbf{x}}, t)$ is a given continuously differentiable function of its arguments. Notice that here $\mathbf{u}(t)$ of Section 8.1 is replaced by $\dot{x}(t)$ as the control vector, so that the only restriction on the control variables is that they be piecewise continuous, and

$$\Omega = \mathbf{R}^n.$$

Historically, there was a great deal of interest in the late seventeenth century in finding the maximum or minimum values of varying quantities. A famous problem, the brachistochrone ($\beta\rho\alpha\chi\iota\sigma\tau\sigma\varsigma$ = shortest, $\chi\rho\acute{o}\nu\sigma\varsigma$ = time) problem, first discussed by Galileo Galilei in his Dialogues (1632, 1638), was to find the shortest time for a small bead to slide without friction under gravity from one given point on a wire to a lower point by varying the shape of the wire. John Bernoulli in 1696 challenged the mathematicians of his day to solve the problem and gave the name brachistochrone to this curve of fastest descent. His brother James and others solved it and found it to be a cycloid.

Legend has it that Queen Dido of Carthage back in the ninth century B.C. intuitively solved a calculus of variations problem. She was promised as large a piece of land as she could enclose within the hide of a bull; she had the bull's hide cut into narrow strips, used the Mediterranean coast as one boundary, and laid the strips in a semicircle along the coast, thus getting the largest possible area. This is an example of an *isoperimetric* problem in the calculus of variations, that is, finding a curve of fixed length (constant perimeter) to enclose the largest or smallest area by varying the shape of the curve.

Euler, a pupil of John Bernoulli, is generally credited with establishing the calculus of variations firmly as a discipline in mathematics. He derived a differential equation, known as the *Euler equation*, which we shall look at shortly. Later Lagrange simplified and generalized Euler's result by introducing a multiplier rule and the equation thus obtained is referred to as the *Euler–Lagrange equation*.

8.2.1 Necessary condition–the Euler equation

Consider a special case of (8.2.1) in which there is only one state variable $x(t)$ and both the initial and terminal points are given, that is,

$$\max_{x(t)} J = \int_{t_0}^{t_f} I\big[x(t), \dot{x}(t), t \big] \, dt, \tag{8.2.2}$$

where $t_0, t_f, x_0 = x(t_0)$, and $x_f = x(t_f)$ are given, and the function I is continuous and twice differentiable with respect to its arguments, and x is a continuous function with piecewise continuous derivative $\dot{x}$. Notice that in (8.2.2) we maximize J over $x(t)$, since $\dot{x}(t)$ is simply dx/dt and determines $x(t)$ up to a constant. Suppose $\bar{x}(t)$ is actually the solution to the problem, that is, $\bar{x}(t)$ is the feasible or admissible trajectory which maximizes the objective functional J. We wish to find out whether it must satisfy any

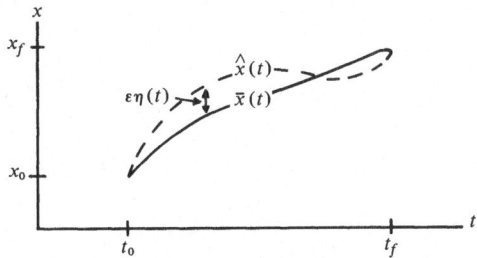

Figure 8.2.1.

necessary condition.[5] Assuming $\bar{x}$ is an interior point of the domain of J, we may consider any neighboring feasible function $\hat{x}(t)$ such that

$$\hat{x}(t) = \bar{x}(t) + \varepsilon\,\eta(t),$$

where ε is some real number and $\eta(t)$ is an arbitrary function with a piecewise continuous derivative $\dot{\eta}$ for which

$$\eta(t_0) = \eta(t_f) = 0. \tag{8.2.3}$$

The function $\hat{x}(t)$ is a variation[6] about the maximizing function $\bar{x}(t)$ and

$$\lim_{\varepsilon \to 0} \hat{x}(t) = \bar{x}(t).$$

Clearly $\hat{x}(t)$ is another feasible trajectory. Its vertical deviation from $\bar{x}(t)$ is given by $\varepsilon\eta(t)$, as shown in Figure 8.2.1. $\bar{x}(t)$ is simply a member of the family of feasible trajectories with $\varepsilon = 0$. Hence the value of the objective functional J for these feasible trajectories is a function of ε:

$$J(\varepsilon) = \int_{t_0}^{t_f} I\left[\bar{x} + \varepsilon\eta, \dot{\bar{x}} + \varepsilon\dot{\eta}, t\right] dt. \tag{8.2.4}$$

By the generalized mean value theorem,

$$I(\bar{x} + \varepsilon\eta, \dot{\bar{x}} + \varepsilon\dot{\eta}, t) = I(\bar{x}, \dot{\bar{x}}, t) + \varepsilon\eta\frac{\partial I}{\partial x} + \varepsilon\dot{\eta}\frac{\partial I}{\partial \dot{x}} + (\text{terms in } \varepsilon^2, \varepsilon^3, \dots)$$

and

$$J(\varepsilon) = \int_{t_0}^{t_f} \left[I(\bar{x}, \dot{\bar{x}}, t) + \varepsilon\eta\frac{\partial I}{\partial x} + \varepsilon\dot{\eta}\frac{\partial I}{\partial \dot{x}} + \dots \right] dt. \tag{8.2.5}$$

Assuming an interior maximum so that $J(\varepsilon)$ is defined over all small ε, the necessary condition for $\bar{x}$ to be the maximizing function for $J(x)$ is that $\varepsilon = 0$ be a critical point of $J(\varepsilon)$, that is, $dJ/d\varepsilon = 0$ at $\varepsilon = 0$. Thus (in view of

---

[5]The reader will find that the situation is analogous to unconstrained classical maximization as discussed in Chapters 1 and 2. There we considered small variations about the solution $f(x)$ and arrived at the necessary condition that the first derivative (or first partial derivatives) must be equal to zero for an interior extremum. Here we shall consider small variations about the maximizing function (the solution trajectory) $x(t)$ and arrive at the Euler equation.

[6]Hence the name calculus of variations.

the fact that I does not depend on ε),

$$\frac{dJ(\varepsilon)}{d\varepsilon}\bigg|_{\varepsilon=0} \int_{t_0}^{t_f}\left(\eta\frac{\partial I}{\partial x}+\dot{\eta}\frac{\partial I}{\partial \dot{x}}\right)dt=\int_{t_0}^{t_f}\eta\frac{\partial I}{\partial x}dt+\int_{t_0}^{t_f}\dot{\eta}\frac{\partial I}{\partial \dot{x}}dt=0. \quad (8.2.6)$$

Integrating the second integral by parts and remembering that

$$\int \dot{\eta}\frac{\partial I}{\partial \dot{x}}dt=\int\frac{\partial I}{\partial \dot{x}}\frac{d\eta}{dt}dt=\int\frac{\partial I}{\partial \dot{x}}d\eta=\eta\frac{\partial I}{\partial \dot{x}}-\int \eta d\left(\frac{\partial I}{\partial \dot{x}}\right)$$

$$=\eta\frac{\partial I}{\partial \dot{x}}-\int \eta\frac{d}{dt}\left(\frac{\partial I}{\partial \dot{x}}\right)dt,$$

we have

$$\frac{dJ(\varepsilon)}{d\varepsilon}=\int_{t_0}^{t_f}\eta\frac{\partial I}{\partial x}dt+\eta\frac{\partial I}{\partial \dot{x}}\bigg|_{t_0}^{t_f}-\int_{t_0}^{t_f}\eta\frac{d}{dt}\left(\frac{\partial I}{\partial \dot{x}}\right)dt=0. \quad (8.2.7)$$

Since $\eta(t_0)=\eta(t_f)=0$, the second term of (8.2.7) drops out and we have

$$\frac{dJ}{d\varepsilon}(0)\doteq\int_{t_0}^{t_f}\eta\left[\frac{\partial I}{\partial x}-\frac{d}{dt}\frac{\partial I}{\partial \dot{x}}\right]dt=0. \quad (8.2.8)$$

(8.2.8) has to hold for all $\eta(t)$ that satisfy (8.2.3), and it can be proved[7] that the expression inside the brackets must then be equal to 0, for otherwise $\eta(t)$ may be chosen to be nonzero at points where this expression is also nonzero, contradicting the requirement that the integral is zero. Hence

$$\frac{\partial I}{\partial x}-\frac{d}{dt}\frac{\partial I}{\partial \dot{x}}=0 \quad (8.2.9)$$

at

$$(x,\dot{x})=\left(\bar{x}(t),\dot{\bar{x}}(t)\right)$$

for $t_0\leqslant t\leqslant t_f$. This is known as the *Euler equation*, which gives a necessary condition for the functional J to attain a maximum at $x(t)=\bar{x}(t)$, $t_0\leqslant t\leqslant t_f$.

Since I is a twice differentiable function of x, $\dot{x}$, and t, and

$$\frac{d}{dt}\frac{\partial I}{\partial \dot{x}}=\frac{\partial}{\partial x}\left(\frac{\partial I}{\partial \dot{x}}\right)\frac{dx}{dt}+\frac{\partial}{\partial \dot{x}}\left(\frac{\partial I}{\partial \dot{x}}\right)\frac{d\dot{x}}{dt}+\frac{\partial}{\partial t}\left(\frac{\partial I}{\partial \dot{x}}\right),$$

(8.2.9) may be rewritten as the following second-order ordinary differential equation:

$$\frac{\partial I}{\partial x}-\left(\frac{\partial^2 I}{\partial x\partial \dot{x}}\dot{x}+\frac{\partial^2 I}{\partial \dot{x}^2}\ddot{x}+\frac{\partial^2 I}{\partial t\partial \dot{x}}\right)=0. \quad (8.2.10)$$

This differential equation, together with the given boundary conditions $x(t_0)=x_0$ and $x(t_f)=x_f$, yields the solution x. Although (8.2.9) or (8.2.10)

---

[7]This is known as the Fundamental Lemma of the calculus of variations. The interested reader may read Takayama (1974), p. 414.

gives only a necessary condition and does not tell whether the solution gives a maximum or minimum or either, practical considerations in many problems will usually help in the determination.

In some special cases, the Euler equation is even easy to solve, as seen below:

(1) The argument $\dot{x}$ does not appear in the intermediate function $I = I[x(t), t]$. In that case (8.2.9) reduces to

$$\frac{\partial I}{\partial x} = 0 \qquad (8.2.11)$$

and this is essentially the first-order condition for an extremum in the classical sense of Chapters 1 and 2.

(2) The argument x does not appear in I. Then (8.2.9) reduces to

$$\frac{d}{dt}\frac{\partial I}{\partial \dot{x}} = 0, \quad i.e., \quad \frac{\partial I}{\partial \dot{x}} = \text{constant.} \qquad (8.2.12)$$

(3) The argument t does not appear explicitly in $I = I[x(t), \dot{x}(t)]$, which is the *automous* case. Then $\partial I / \partial t = 0$ and

$$\frac{dI}{dt} = \frac{\partial I}{\partial x}\dot{x} + \frac{\partial I}{\partial \dot{x}}\ddot{x} + 0 = \frac{d}{dt}\left(\frac{\partial I}{\partial \dot{x}}\right)\dot{x} + \frac{\partial I}{\partial \dot{x}}\ddot{x} = \frac{d}{dt}\left[\dot{x}\frac{\partial I}{\partial \dot{x}}\right]$$

by making use of the Euler equation, which states that

$$\frac{\partial I}{\partial x} = \frac{d}{dt}\left(\frac{\partial I}{\partial \dot{x}}\right)$$

at an extremum. The two functions I and $\dot{x}\,\partial I / \partial \dot{x}$ have the same derivative with respect to t, thus implying that they differ only by a constant, that is,

$$I = \dot{x}\frac{\partial I}{\partial \dot{x}} + \text{constant,}$$

which is a first-order differential equation that can be solved fairly easily.

Example 1. As an illustration of how the Euler equation is used, let us consider an extremely simple problem, maximizing

$$J = \int_0^1 -\sqrt{1 + \left(\frac{dx}{dt}\right)^2}\, dt,$$

where we specify that $t_0 = 0$, $t_f = 1$, $x_0 = 1$, $x_f = 2$. The intermediate function is $-\sqrt{1 + \dot{x}^2}$.

Since x does not appear in the intermediate function I, we use (8.2.12) to get

$$\frac{\partial I}{\partial \dot{x}} = \frac{\dot{x}}{\sqrt{1 + \dot{x}^2}} = \text{constant,}$$

which means that $\dot{x}$ is a constant. This in turn yields the linear solution

$$x(t) = c_1 t + c_2.$$

The constants c_1 and c_2 are obtained by substituting in the prescribed boundary values as follows:

$$1 = c_1 \cdot 0 + c_2;$$
$$2 = c_1 \cdot 1 + c_2.$$

Hence $c_1 = c_2 = 1$, or $x = t + 1$. What we just did was to show that the curve giving the minimum distance between the points $(0, 1)$ and $(1, 2)$ is the line $x = t + 1$ (remembering that maximizing the negative of a functional is equivalent to minimizing it). $\sqrt{1 + \dot{x}^2} \; dt$ or $\sqrt{(dt)^2 + (dx)^2}$ is simply ds, the differential of arc length. $\qquad\qquad\qquad\qquad\qquad\qquad\qquad\square$

The Euler equation, (8.2.9), was derived using only one state variable $x(t)$ and assuming that both end points (t_0, x_0) and (t_f, x_f) are given. A similar result can be obtained when we relax these assumptions. If we have a multidimensional case with a vector $\mathbf{x}(t)$ of n state variables, the n Euler equations read:

$$\frac{\partial I}{\partial \mathbf{x}} - \frac{d}{dt}\frac{\partial I}{\partial \dot{\mathbf{x}}} = 0, \quad i.e., \quad \nabla_{\mathbf{x}} I - \frac{d}{dt}(\nabla_{\dot{\mathbf{x}}} I) = 0, \qquad (8.2.13)$$

where

$$\nabla_{\mathbf{x}} I = \begin{bmatrix} \dfrac{\partial I}{\partial x_1} \\[2mm] \dfrac{\partial I}{\partial x_2} \\[1mm] \vdots \\[1mm] \dfrac{\partial I}{\partial x_n} \end{bmatrix}, \quad \nabla_{\dot{\mathbf{x}}} I = \begin{bmatrix} \dfrac{\partial I}{\partial \dot{x}_1} \\[2mm] \dfrac{\partial I}{\partial \dot{x}_2} \\[1mm] \vdots \\[1mm] \dfrac{\partial I}{\partial \dot{x}_n} \end{bmatrix}, \quad \text{and} \quad \frac{d}{dt}(\nabla_{\dot{\mathbf{x}}} I) = \begin{bmatrix} \dfrac{d}{dt}\dfrac{\partial I}{\partial \dot{x}_1} \\[2mm] \dfrac{d}{dt}\dfrac{\partial I}{\partial \dot{x}_2} \\[1mm] \vdots \\[1mm] \dfrac{d}{dt}\dfrac{\partial I}{\partial \dot{x}_n} \end{bmatrix}.$$

Example 2:[8] Consider a one-sector economy in which a single homogeneous good is produced by two factors of production, labor and capital. Let the states of labor and capital over time be $x_1(t)$ and $x_2(t)$ respectively, and assume that the time rate of growth of capital or rate of savings, $dx_2/dt = \dot{x}_2$, can be controlled. The total product or income of that society is a function of x_1 and x_2, namely, $f(x_1, x_2)$. Consumption $c(t)$ is then

$$c(t) = f(x_1, x_2) - \dot{x}_2. \qquad (8.2.14)$$

Assuming that the utility of the society's consumption is a function of the amount of consumption alone, $u = u[c(t)]$, and that the disutility of work is $v = v(x_1)$, the total utility of the society over the period of time from t_0 to t_f

[8]This example is from Ramsey, F.P., 1928. A mathematical theory of saving, *Economic Journal*, pp. 543–549. It is a much simplified version of Hadley and Kemp (1971), pp. 50 on, and Allen (1938), pp. 537–40.

is

$$J = \int_{t_0}^{t_f} \{ u[c(t)] - v[x_1(t)] \} \, dt$$

$$= \int_{t_0}^{t_f} \{ u[f(x_1, x_2) - \dot{x}_2] - v(x_1) \} \, dt$$

$$= \int_{t_0}^{t_f} I(x_1, x_2, \dot{x}_2) \, dt,$$

where $I(x_1, x_2, \dot{x}_2) = u[f(x_1, x_2) - \dot{x}_2] - v(x_1)$. It is clear from the above that there are two state variables x_1 and x_2 and that we wish to find that time path $\dot{x}_2$ of the stock of capital which maximizes the utility functional J. By the necessary Euler equations given by (8.2.13) we must have

$$\frac{\partial I}{\partial x_1} = \frac{d}{dt} \left(\frac{\partial I}{\partial \dot{x}_1} \right) \quad \text{and} \quad \frac{\partial I}{\partial x_2} = \frac{d}{dt} \left(\frac{\partial I}{\partial \dot{x}_2} \right). \qquad (8.2.15)$$

Now, $\partial I / \partial \dot{x}_1 = 0$; thus

$$\frac{\partial I}{\partial x_1} = \frac{du}{dc} \frac{\partial f}{\partial x_1} - \frac{dv}{dx_1} = 0,$$

so that

$$\frac{\partial f}{\partial x_1} = \frac{dv}{dx_1} \bigg/ \frac{du}{dc}. \qquad (8.2.16)$$

The second condition in (8.2.15) gives

$$\frac{du}{dc} \frac{\partial f}{\partial x_2} = \frac{d}{dt} \left(\frac{\partial I}{\partial \dot{x}_2} \right) = \frac{d}{dt} \left[\frac{du}{dc} (-1) \right] = \frac{d}{dt} \left(-\frac{du}{dc} \right)$$

so that

$$\frac{\partial f}{\partial x_2} = \left(-1 \bigg/ \frac{du}{dc} \right) \frac{d}{dt} \left(\frac{du}{dc} \right). \qquad (8.2.17)$$

If we assume that the amounts of labor and capital at the initial time t_0 are known and that the society specifies the amounts for these at terminal time t_f, then these boundary conditions, together with the equations given by (8.2.14), (8.2.16), and (8.2.17), can be used to solve for $c(t)$, $x_1(t)$, and $x_2(t)$ (also assuming that the utility and disutility functions u and v are measurable and given).

The partial derivatives $\partial f / \partial x_1$ and $\partial f / \partial x_2$ may be interpreted to represent the rate of wages (or marginal product of labor) and rate of interest (marginal product of capital) respectively. It is of interest to note from (8.2.16) that the rate of wages is equal to the ratio of the marginal disutility of work dv/dx_1 to the marginal utility of consumption du/dc. And from (8.2.17) we note that the rate of interest is equal to the proportional rate of decrease of the marginal utility of consumption over time, the proportionality factor being the reciprocal of the marginal utility of consumption. □

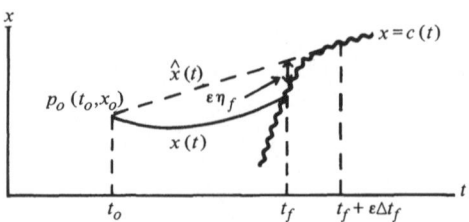

Figure 8.2.2.

8.2.2 Transversality condition

For simplicity of discussion, we have confined ourselves to problems with given fixed end points (t_0, x_0) and (t_f, x_f). In more general situations, the end points may vary over some specified loci, and one wishes to establish necessary conditions, if any, which the solution curve must satisfy to be the best among neighboring curves with neighboring end points, thus yielding, for the functional to be maximized (or minimized), an extremum. Such necessary conditions are called *transversality conditions*.

Let us consider only a problem with one state variable x and with the initial end point (t_0, x_0) given but the terminal point lying on some curve $x = c(t)$. We shall sketch the derivation of the transversality condition for this problem briefly.[9]

We proceed as we did in subsection 8.2.1. To avoid notational clutter, we assume that $x(t)$ is the solution (instead of $\bar{x}$ as previously) and we will omit the arguments of the functions whenever there should be no confusion. Let $\hat{x}$ be a neighboring function where

$$\hat{x}(t) = x(t) + \varepsilon \eta(t),$$

and $\eta_0 = \eta(t_0) = 0$ but $\eta_f = \eta(t_f)$ is not specified. As done previously, we form the functional

$$J(\varepsilon) = \int_{t_0}^{t_f + \varepsilon \Delta t_f} I(x + \varepsilon \eta, \dot{x} + \varepsilon \dot{\eta}, t) \, dt$$

and see that a necessary condition for J to be an extremum is that

$$\frac{dJ(\varepsilon)}{d\varepsilon} = 0 \quad \text{at } \varepsilon = 0.$$

It can be shown, by reasoning similar to that employed in deriving (8.2.6), that

$$\frac{dJ(\varepsilon)}{d\varepsilon}\Big|_{\varepsilon=0} = \int_{t_0}^{t_f} \left(\eta \frac{\partial I}{\partial x} + \dot{\eta} \frac{\partial I}{\partial \dot{x}} \right) dt + I(t_f) \Delta t_f. \qquad (8.2.18)$$

---

[9]The interested reader may consult D.R. Smith (1974), pp. 149–164, for a fuller discussion.

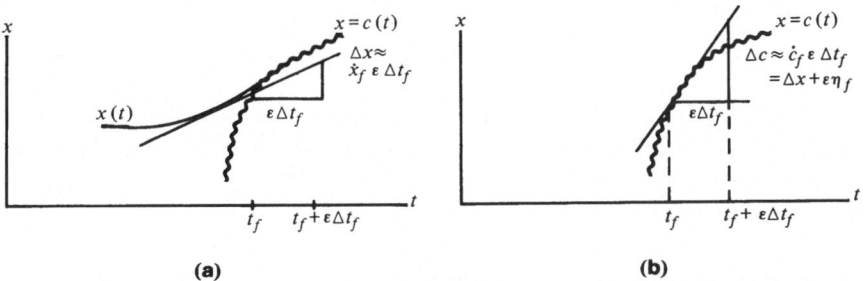

Figure 8.2.3

Again, after integrating by parts the second term of the integral in (8.2.18), we conclude that the necessary condition is that

$$\int_{t_0}^{t_f} \eta \left[\frac{\partial I}{\partial x} - \frac{d}{dt}\left(\frac{\partial I}{\partial \dot{x}} \right) \right] dt + \eta_f \frac{\partial I}{\partial \dot{x}}\Big|_{t_f} - \eta_0 \frac{\partial I}{\partial \dot{x}}\Big|_{t_0} + I(t_f)\Delta t_f = 0. \quad (8.2.19)$$

Let us now look at Figures 8.2.2 and 8.2.3 (a) and (b). In Figure 8.2.3 (a), we show the tangent to the curve $x(t)$ at the point (t_f, x_f), and in Figure 8.2.3 (b) we show the tangent to the curve $c(t)$ at the same point. From these figures, we see that $x(t_f) = c(t_f)$ and

$$\dot{x}_f \varepsilon \Delta t_f + \varepsilon \eta_f = \dot{c}_f \varepsilon \Delta t_f,$$

so that

$$\eta_f = (\dot{c}_f - \dot{x}_f)\Delta t_f \quad (8.2.20)$$

where $\dot{c}_f$ and $\dot{x}_f$ are the derivatives dc/dt and dx/dt respectively evaluated at $t = t_f$.

Substituting (8.2.20) into (8.2.19) yields

$$\int_{t_0}^{t_f} \eta \left[\frac{\partial I}{\partial x} - \frac{d}{dt}\left(\frac{\partial I}{\partial \dot{x}} \right) \right] dt + \left[I(t_f) + (\dot{c}_f - \dot{x}_f)\frac{\partial I(t_f)}{\partial \dot{x}} \right]\Delta t_f - \eta_0 \frac{\partial I}{\partial \dot{x}}\Big|_{t_0} = 0.$$

$$(8.2.21)$$

Since $\partial I/\partial \dot{x}$ is finite at $t = t_0$ and $\eta_0 = \eta(t_0) = 0$, the last term of (8.2.21) drops out. Δt_f is arbitrary, and may be set equal to zero, which leads to the conditions

$$\int_{t_0}^{t_f} \eta \left[\frac{\partial I}{\partial x} - \frac{d}{dt}\left(\frac{\partial I}{\partial \dot{x}} \right) \right] dt = 0 \quad (8.2.22)$$

and

$$I(t_f) + (\dot{c}_f - \dot{x}_f)\frac{\partial I}{\partial x}\Big|_{t_f} = 0. \quad (8.2.23)$$

(8.2.22) must hold for all η satisfying $\eta(t_0) = 0$; hence we have the Euler equation again,

$$\frac{\partial I}{\partial x} - \frac{d}{dt}\left(\frac{\partial I}{\partial \dot{x}}\right) = 0.$$

(8.2.23) is the transversality condition, which, together with the initial boundary condition t_0 and $x(t_0)$ and the Euler equation, hopefully determines the solution curve $x(t)$. We say hopefully because these conditions are only necessary, not sufficient.

8.2.3 Constraints and Lagrange multipliers

Certain constraints can be handled quite easily in a classical calculus of variations problem. The simplest type involves a constraint which appears as an integral, say,

$$\int_{t_0}^{t_f} g(x, \dot{x}, t)\, dt = \text{constant},$$

where g is a continuously differentiable function. Introduce the Lagrange multiplier λ as in the previous chapters, and maximize the Lagrangian functional

$$J' = \int_{t_0}^{t_f}\left[I(x, \dot{x}, t) + \lambda g(x, \dot{x}, t)\right] dt$$

subject to the same boundary conditions. It can be shown that the associated Euler condition is

$$\frac{\partial(I+\lambda g)}{\partial x} - \frac{d}{dt}\left[\frac{\partial}{\partial \dot{x}}(I+\lambda g)\right] = 0 \qquad (8.2.24)$$

by using the same line of reasoning as used in Subsection 8.2.1.

If the constraint is not an integral but is of the form

$$g(x, \dot{x}, t) \leqslant b,$$

where b is an $r \times 1$ vector of constants and g represents a vector of r functions $g_1(x_1, \ldots, x_n, \dot{x}_1, \ldots, \dot{x}_n, t), \ldots, g_r(x_1, \ldots, x_n, \dot{x}_1, \ldots, \dot{x}_n, t)$, we introduce a $1 \times r$ Lagrange multiplier vector $\lambda(t) = (\lambda_1(t), \ldots, \lambda_r(t))$ and consider the functional

$$J'(x, \lambda) = \int_{t_0}^{t_f}\left\{ I(x, \dot{x}, t) + \lambda\left[b - g(x, \dot{x}, t)\right]\right\} dt. \qquad (8.2.25)$$

We take the λ's to be functions of t since the constraints must be satisfied for all t, $t_0 \leqslant t \leqslant t_f$. The problem is analogous to that found in the static case of the earlier chapters. The Lagrangian $L = I + \lambda(b - g)$ here is the integrand of (8.2.25), and we wish to find the solution vector $x(t)$ which maximizes J' and the vector λ which minimizes it, subject to the necessary conditions

$$\frac{\partial L}{\partial x} - \frac{d}{dt}\frac{\partial L}{\partial \dot{x}} = 0,$$

$$g(x, \dot{x}, t) \leqslant b, \qquad (8.2.26)$$

$$\lambda\left[b - g(x, \dot{x}, t)\right] = 0 \quad \text{and} \quad \lambda \geqslant 0.$$

The first set of equations are known as the *Euler–Lagrange equations*[10] and there are n of them, one for each x_i. Then there are r inequalities, which are the given constraints. The remaining conditions are simply the Kuhn–Tucker conditions of Chapter 7.

Example. Let us find the necessary condition for the vector $\mathbf{x}(t) = (x_1(t), x_2(t))^T$ to extremize the functional

$$J = \int_0^{t_f} \left(\tfrac{1}{2} x_1^2 + \tfrac{1}{2} x_2^2 \right) dt,$$

with

$$x_1(0) = 0, \qquad x_2(0) = 1,$$

subject to

$$\dot{x}_1(t) = x_2(t).$$

In this problem we have taken $t_0 = 0$. There is only one constraint, $x_2 - \dot{x}_1 = 0$, so we introduce only one Lagrange multiplier λ and form the Lagrangian

$$L = \tfrac{1}{2} x_1^2 + \tfrac{1}{2} x_2^2 + \lambda(x_2 - \dot{x}_1).$$

By the Euler–Lagrange equations, the solution vector must satisfy:

$$\frac{\partial L}{\partial x_1} - \frac{d}{dt} \frac{\partial L}{\partial \dot{x}_1} = x_1 - \frac{d}{dt}(-\lambda) = x_1 + \dot{\lambda} = 0, \quad i.e., \quad x_1 = -\dot{\lambda};$$

$$\frac{\partial L}{\partial x_2} - \frac{d}{dt} \frac{\partial L}{\partial \dot{x}_2} = x_2 + \lambda - \frac{d}{dt}(0) = x_2 + \lambda = 0, \quad i.e., \quad x_2 = -\lambda.$$

These, coupled with the constraint $x_2 = \dot{x}_1$, yield the conditions:

$$x_1 = \dot{x}_2;$$
$$x_2 = \dot{x}_1.$$

These are two second-order differential equations,

$$x_1 = \ddot{x}_1 = \frac{d^2 x_1}{dt^2} \quad \text{and} \quad x_2 = \ddot{x}_2 = \frac{d^2 x_2}{dt^2},$$

and their solutions are

$$x_1 = Ae^t + Be^{-t},$$
$$x_2 = Ae^t - Be^{-t}.$$

From the initial boundary condition,

$$x_1(0) = 0 \quad \text{and} \quad x_2(0) = \dot{x}_1(0) = 1,$$

we have $A + B = 0$ and $A - B = 1$, or $A = \tfrac{1}{2}$, $B = -\tfrac{1}{2}$. Therefore

$$x_1 = \tfrac{1}{2} e^t - \tfrac{1}{2} e^{-t} \quad \text{and} \quad x_2 = \tfrac{1}{2} e^t + \tfrac{1}{2} e^{-t}.$$

---

[10]The Euler equation of (8.2.9) does not involve any multiplier λ or constraints, whereas in (8.2.26) we take the partial derivatives of the Lagrangian which does contain the constraint **g**.

8.2.4 Higher order derivatives

The integrand I may be a function of higher order derivatives of $\mathbf{x}$ as well as of the other three arguments t, $\mathbf{x}$, and $\dot{\mathbf{x}}$. Taking the case of a single state variable as illustration, we may wish to find the trajectory $x(t)$ which maximizes the functional

$$J(x) = \int_{t_0}^{t_f} I(x, \dot{x}, \ddot{x}, t)\, dt. \tag{8.2.27}$$

This may be transformed into a problem of the type we have studied by letting

$$x = x_1, \qquad \dot{x}_1 = x_2, \tag{8.2.28}$$

so that (8.2.27) may be written as

$$J = \int_{t_0}^{t_f} I(x_1, x_2, \dot{x}_2, t)\, dt$$

subject to (8.2.28).

Actually, it is just as easy not to use any such transformation. It can be shown that if the functional is

$$J = \int_{t_0}^{t_f} I\left(x, \dot{x}, \ddot{x}, \ldots, \frac{d^k x}{dt^k}, t\right) dt,$$

a necessary condition for a solution trajectory $x(t)$ to extremize J is

$$\frac{\partial I}{\partial x} + \sum (-1)^i \frac{d^i}{dt^i}\left(\partial I / \partial\left(d^i x / dt^i\right)\right) = 0, \tag{8.2.29}$$

which is known as the Euler–Poisson equation.

8.2.5 Other necessary conditions

The Euler equations given by (8.2.9), (8.2.26), or (8.2.29) are by no means the only necessary conditions that a solution trajectory must satisfy. We have already mentioned the boundary and transversality conditions. In addition to these, there are some others. One is the *Weierstrass condition*, which states that in the problem of maximizing

$$J = \int_{t_0}^{t_f} I(\mathbf{x}, \dot{\mathbf{x}}, t)\, dt$$

subject to $g^i(\mathbf{x}, \dot{\mathbf{x}}, t) = 0$ $(i = 1, 2, \ldots, r)$ and the boundary conditions $(t_0, \mathbf{x}_0, t_f, \mathbf{x}_f) \in T$ (T a specified set), and where the Lagrangian function L is

$$L(\mathbf{x}, \dot{\mathbf{x}}, t, \lambda_1, \ldots, \lambda_r) = I(\mathbf{x}, \dot{\mathbf{x}}, t) + \sum_{i=1}^{r} \lambda_i g^i(\mathbf{x}, \dot{\mathbf{x}}, t),$$

if $\bar{\mathbf{x}}(t)$ is the solution trajectory and $\mathbf{x}(t)$ is any other feasible trajectory, then

$$L(\bar{\mathbf{x}}, \dot{\mathbf{x}}, t, \lambda_1, \ldots, \lambda_r) - \sum_{i=1}^{n} \dot{x}_i \frac{\partial I(\bar{\mathbf{x}}, \dot{\bar{\mathbf{x}}}, t, \lambda_1, \ldots, \lambda_r)}{\partial \dot{x}_i}$$

$$\leqslant L(\bar{\mathbf{x}}, \dot{\bar{\mathbf{x}}}, t, \lambda_1, \ldots, \lambda_r) - \sum_{i=1}^{n} \dot{\bar{x}}_i \frac{\partial I}{\partial \dot{x}_i}(\bar{\mathbf{x}}, \dot{\bar{\mathbf{x}}}, t, \lambda_1, \ldots, \lambda_r).$$

(8.2.30)

Where there is only one state variable and no constraint, the Weierstrass condition reduces to

$$I(\bar{x}, \dot{x}, t) - \dot{x} \frac{\partial I(\bar{x}, \dot{\bar{x}}, t)}{\partial \dot{x}} \leqslant I(\bar{x}, \dot{\bar{x}}, t) - \dot{\bar{x}} \frac{\partial I(\bar{x}, \dot{\bar{x}}, t)}{\partial \dot{x}}.$$

There are other necessary conditions, but the Euler equations are probably the most useful in helping us find the solution.

We have not mentioned sufficient conditions because they are much more complicated and involve the concept of convexity.

8.3 The maximum principle (modern calculus of variations)

One of the major limitations of the classical calculus of variations approach is that the control variable $\mathbf{u}(t)$ is $\dot{\mathbf{x}}(t)$, the derivative of $\mathbf{x}$ with respect to t, and that, except for the restriction that $\dot{\mathbf{x}}$ be piecewise continuous, the problem does not lend itself easily to many types of restrictions on the control variables. The method discussed in Subsection 8.2.3 may not be applicable. For instance, a situation which is often encountered in control problems is one in which we require the control set Ω to be a cube in Euclidean r-space as defined by

$$|u_j(t)| \leqslant 1, \qquad j = 1, \ldots, r.$$

It usually turns out that the optimal control for such problems is the point defined by

$$|u_j(t)| = 1, \qquad j = 1, \ldots, r.$$

That is, the point will move from one vertex of Ω to another according to the optimal control rule worked out as a solution to the particular problem under consideration. The classical calculus variations approach is unable to handle such a problem. Notice, moreover, that $\mathbf{u}(t)$ here is not $\dot{\mathbf{x}}(t)$.

In a series of papers published in the late 1950s, the Russian mathematician L.S. Pontryagin and his coworkers presented what is now known as the *Pontryagin maximum principle*. This marked the beginning of modern control theory. The maximum principle is in essence a generalization of the classical calculus of variations, and makes use of a concept known as the Hamiltonian, which was developed in solving problems in mechanics. We shall present a simplified version of the maximum principle, using a vector

$\mathbf{x}(t)$ of n state variables $x_i(t)$ and a single control $u(t)$. It is an easy extension to problems with several control variables.

The problem is the same as that stated in (8.1.3), which we repeat:

$$J = \int_{t_0}^{t_f} I(\mathbf{x}, u, t) \, dt + F(\mathbf{x}_f, t_f)$$

subject to $\dot{x}_i = dx_i/dt = f_i(\mathbf{x}, u, t)$, $i = 1, \ldots, n$, where $u(t)$ is some piecewise continuous function restricted to belong to some control set $\mathcal{U}$, t_0 and $\mathbf{x}_0 = \mathbf{x}(t_0)$ are given, t_f and $\mathbf{x}_f = \mathbf{x}(t_f)$ are either given or required to satisfy some specified relation $T(\mathbf{x}, t) = 0$, the function f is continuous in its arguments and continuously differentiable with respect to $\mathbf{x}$, and the functions I and F are continuously differentiable with respect to their arguments. Of all the admissible controls $u(t)$ (meaning $u \in \mathcal{U}$), and of all the corresponding solutions $\mathbf{x}(t)$ (which may be many for the same $u \in \mathcal{U}$), we seek the pair $(\mathbf{x}(t), u(t))$ which will give J the largest possible value.

We first introduce a nonzero vector function $\boldsymbol{\lambda}(t) = (\lambda_1(t), \ldots, \lambda_n(t))^T$ whose coordinates λ_i we shall call *costate* variables or *auxiliary* variables, and form the expression

$$H(\mathbf{x}, u, \boldsymbol{\lambda}, t) = I(\mathbf{x}, u, t) + \sum_{i=1}^{n} \lambda_i(t) f_i(\mathbf{x}, u, t), \qquad (8.3.1)$$

which is called the *Hamiltonian function*. This $\boldsymbol{\lambda}(t)$ should remind one of the Lagrange multiplier in classical optimization and in Subsection 8.2.3, and indeed it is the dynamic equivalent of the former.

The *maximum principle*[11] states that if the admissible control $\bar{u}(t)$ and its resulting state variable $\bar{\mathbf{x}}(t)$ are optimal for the functional J, then there must necessarily exist a nonzero function $\boldsymbol{\lambda}(t)$ corresponding to $\bar{u}$ and $\bar{\mathbf{x}}$ such that for all t, $t_0 \leqslant t \leqslant t_f$:

(1) $\dfrac{dx_i}{dt} = \dfrac{\partial H}{\partial \lambda_i}$, $i = 1, \ldots, n$, $\qquad (8.3.2)$

(2) $\dfrac{d\lambda_i}{dt} = -\dfrac{\partial H}{\partial x_i}$, $i = 1, \ldots, n$; $\qquad (8.3.3)$

(3) H attains its maximum at $\bar{u}(t)$, that is,

$$H(\bar{\mathbf{x}}, \bar{u}, \boldsymbol{\lambda}, t) \geqslant H(\bar{\mathbf{x}}, u, \boldsymbol{\lambda}, t)$$

  for all $u \in \mathcal{U}$; $\qquad (8.3.4)$

(4) $\dfrac{\partial F}{\partial x_i}\bigg|_{\mathbf{x}_f} = \lambda_i(t_f)$, $i = 1, \ldots, n$,   and $\mathbf{x}_0 = \mathbf{x}(t_0)$. $\qquad (8.3.5)$

---

[11]The interested reader will be pleased to know that the English translation of Pontryagin *et al.* (1962) is a very readable book and gives a formal treatment of the maximum principle. In it, the authors seek the control $\mathbf{u} = \mathbf{u}(t)$ that yields the least possible value for the functional $J = \int_{t_0}^{t_f} I(\mathbf{x}, \mathbf{u}, t) \, dt$, and the Hamiltonian contains the term $\psi_0(t)I$, where $\psi_0(t) < 0$. Namely, the Hamiltonian is $H(\mathbf{x}, \mathbf{u}, \psi_0, \psi, t) = \psi_0 I + \sum_{i=1}^{n} \psi_i f_i$. Here, we seek to maximize J, and the $\psi_0(t)$ in our Hamiltonian should be nonnegative. In most control problems, however, we may assume $\psi_0(t)$ to be positive, in which case we can normalize the $n+1$ costate variables $\psi_i(t)$, $i = 0, 1, \ldots, n$, by dividing them by ψ_0 to get $\lambda_0 = 1$ and $\lambda_i = \psi_i/\psi_0$, $i = 1, \ldots, n$. Different authors use different symbols for the costate variables, such as ψ, z, y, λ. We use λ to emphasize its close relationship with the Lagrange multipliers of the previous chapters.

It is understood that all the derivatives above are evaluated at $u = \bar{u}(t)$.

The n equations in (8.3.2) are simply the equations of motion $\dot{x}_i = f_i$. These, together with the n equations in (8.3.3) are called the *Hamiltonian canonical equations*. If, in (8.3.4), H attains its maximum at an interior point $\bar{u}$, then we have also

$$\frac{\partial H}{\partial u} = 0. \tag{8.3.6}$$

The $2n$ equations in (8.3.5) are the boundary or transversality conditions.

Example 1.[12] We shall look at a very simple time optimal problem to see how the maximum principle technique works. We wish to minimize the time required to move a process from a given initial state to, say, the origin $O = (0,0)$, *i.e.*,

$$\min_u \int_{t_0}^{t_f} dt.$$

Let there be two state variables, x_1 and x_2, with

$$\dot{x}_1 = f_1(x_1, x_2, u, t) = x_2, \qquad \dot{x}_2 = f_2(x_1, x_2, u, t) = u,$$

where u is the control constrained by

$$|u(t)| \leqslant 1.$$

The problem is the same as

$$\max_u J = \int_{t_0}^{t_f} -1 \, dt,$$

and I is the constant function with $I(x_1, x_2, u, t) = -1$. The Hamiltonian is

$$H = -1 + \lambda_1 x_2 + \lambda_2 u \tag{8.3.7}$$

and, by the canonical equations given by (8.3.3),

$$\frac{d\lambda_1}{dt} = -\frac{\partial H}{\partial x_1} = 0, \qquad \frac{d\lambda_2}{dt} = -\frac{\partial H}{\partial x_2} = -\lambda_1.$$

Integrating, we get $\lambda_1 = c_1$, $\lambda_2 = c_2 - c_1 t$, where the c's are constants. According to the maximum principle, in order to maximize the functional J with respect to u, it is necessary that the Hamiltonian function H be maximized with respect to u. Looking at (8.3.7) and considering the constraint on u, it is easy to see that H is maximized if u has the same sign as λ_2 so as to render that term positive. Therefore we take

$$u(t) = \begin{cases} 1 & \text{if } c_2 - c_1 t \geqslant 0 \\ -1 & \text{if } c_2 - c_1 t < 0 \end{cases}$$

for $t_0 \leqslant t \leqslant t_f$. $\dot{x}_2 = 1$ during the time interval when $u = 1$ so that we have $x_2 = t + k_2$; from $\dot{x}_1 = x_2$ it follows that

$$x_1 = \frac{t^2}{2} + k_2 t + k_1 = \tfrac{1}{2}(t + k_2)^2 + \left(k_1 - \frac{k_2^2}{2}\right),$$

[12]This example is from Pontrayagin *et al.* (1962), pp. 23–27.

giving $x_1 = \frac{1}{2}x_2^2 + k$, where the k's are constants. Similarly, during the time interval when $u = -1$, we get

$$x_2 = -t + \text{constant}, \qquad x_1 = -\frac{1}{2}x_2^2 + \text{constant}.$$

Thus we have two families of parabolas, some of which are shown in the phase diagram of Figure 8.3.1 below. The phase points (x_1, x_2) move in the directions of the arrows (upwards when $dx_2/dt = u = 1$ and downwards when $dx_2/dt = u = -1$). From the phase diagram, we see that the shortest time path to reach the origin $(0,0)$ from any initial point is one of the two paths shown in Figures 8.3.2(a) and 8.3.2(b). For instance, if the process was initiated at point A in Figure 8.3.1 (*i.e.*, if $\mathbf{x}_0 = A$ and $u = 1$), the phase or state trajectory moves up the arc till it hits B on the arc BO (which is a section of the parabola $x_1 = -\frac{1}{2}x_2^2$, with $u = -1$). At that point, u switches to -1 and the state trajectory moves down BO till it hits the origin. On the other hand, if the process was initiated at C, then the state trajectory moves down ($u = -1$) till it hits D on the arc OD (which is a section of the parabola $x_1 = \frac{1}{2}x_2^2$ with $u = +1$). At D, it moves up to hit the origin. In general, if the initial point P is below the BOD curve (where the upper portion BO is a section of the parabola $x_1 = -\frac{1}{2}x_2^2$ and the lower portion OD is a section of $x_1 = \frac{1}{2}x_2^2$), the optimal time path should be controlled by the optimal control $u = +1$ till it hits the curve BO, at which point the control should switch to $u = -1$. The optimal time path is the darkened curve in Figure 8.3.2(a). It is the other way round if the initial point P is above the curve BOD; the optimal course of action should be $u = -1$ first, followed by $u = +1$, as shown in Figure 8.3.2(b). A solution in which the control switches from one value to the other between these two values is called the *bang-bang control*. ☐

   In general, the maximum principle technique is as follows: Use the canonical equations given by (8.3.2) and (8.3.3) and the inequality given by (8.3.4) to determine all pairs $\bar{u}(t)$ and $\bar{x}(t)$ up to general form. We then need the transversality conditions to determine the specific optimal solu-

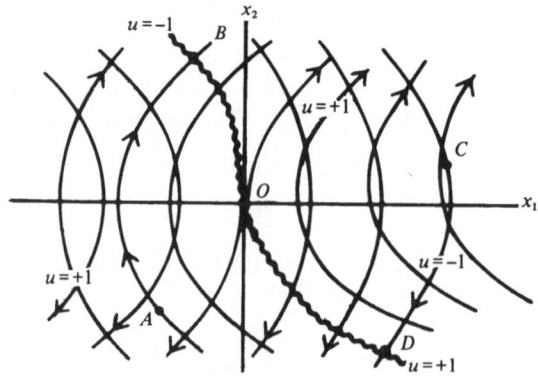

Figure 8.3.1.

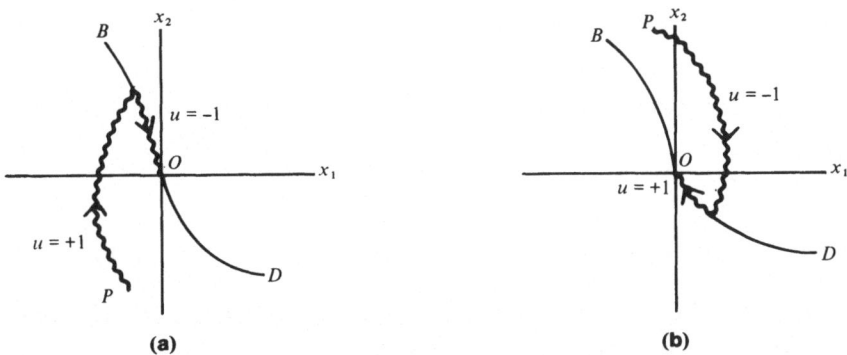

Figure 8.3.2

tion pair. Unfortunately, the transversality conditions in (8.3.5) are such that n of them (for $\mathbf{x}_0$) are given at initial time and the other n (for λ_i) at terminal time, leading to a two-point boundary problem which is usually hard to solve, especially if the dimension of the state vector is high. In this event, most methods use either the trial-and-error approach by choosing $\lambda_i(t_0)$ to satisfy the equations given by (8.3.2), (8.3.3), and (8.3.4), or use numerical approximation on high-speed computers. These methods are not always satisfactory. Although one may not find it feasible to get the exact solution, one can often have a good idea of the nature or structure of the solution by interpreting the results with the use of the phase diagram as we did in Example 1 above and which we will do again in Example 2, which follows.

Example 2.[13] Let there be a one-sector economy with zero population growth and zero technical progress and production under constant returns to scale. Let the population be N, and let $k(t)$ be the capital stock at date t. Then $x(t) = k(t)/N$ is capital per capita. Since N is constant over time, we normalize it so as to be equal to 1 (measured in thousands or millions of persons). Denote per capita consumption by $u(t)$ and the utility enjoyed by an individual consuming at rate u by $\phi(u)$. The production function of the economy is $Y(t) = Ng(x(t)) = g(x)$, and gross investment is equal to output minus consumption, or $Y - Nu = g(x) - u$. Assume that the unit depreciation rate of capital is δ, so that the total depreciation rate of capital stock is δx. Therefore $\dot{x}$, net capital accumulation, which is equal to gross investment minus capital depreciation, is

$$\dot{x} = g(x) - u - \delta x. \tag{8.3.8}$$

Let ρ be the social rate of time preference, so that the utility at time 0 of

---

[13]This is a grossly simplified way of looking at optimal economic growth and is based mainly on R. Dorfman (1969), pp. 824–27. See Arrow (1968) and, in particular, Intriligator (1971), Chapter 16, for a fuller treatment of the topic.

the consumption achieved at time t is

$$e^{-\rho t}\phi(u).$$

Our objective is to maximize the sum of the utilities between times 0 and T, i.e.,

$$\max_{u(t)} J = \int_0^T e^{-\rho t}\phi(u)\,dt,$$

where the process is characterized by (8.3.8). Here $I = e^{-\rho t}\phi(u)$ and $\dot{x} = f(u,x,t) = g(x) - u - \delta x$.

Applying the maximum principle and assuming that our Hamiltonian H may be maximized by an interior $u(t)$, we have

$$H(x,u,\lambda,t) = e^{-\rho t}\phi(u) + \lambda(g(x) - u - \delta x), \qquad (8.3.9)$$

$$\dot{\lambda} = \frac{d\lambda}{dt} = -\frac{\partial H}{\partial x} = -\lambda(g'(x) - \delta) \quad \text{(by (8.3.3))} \quad \text{so that} \quad g'(x) = \delta - \frac{\dot{\lambda}}{\lambda},$$

$$(8.3.10)$$

$$\frac{\partial H}{\partial u} = e^{-\rho t}\phi'(u) - \lambda = 0 \quad \text{(by (8.3.6))} \quad \text{so that} \quad \lambda = e^{-\rho t}\phi'(u), \qquad (8.3.11)$$

where $g'(x) = dg/dx$ and $\phi'(u) = d\phi/du$. If we differentiate (8.3.11) with respect to t and divide the result by λ, we get

$$\frac{\dot{\lambda}}{\lambda} = -\rho + \frac{\phi''(u)}{\phi'(u)}\frac{du}{dt},$$

and, substituting into (8.3.10),

$$g'(x) = \rho + \delta - \frac{\phi''(u)}{\phi'(u)}\dot{u} \quad \text{so that} \quad \dot{u} = \frac{\phi'(u)}{\phi''(u)}\left[\rho + \delta - g'(x)\right]. \quad (8.3.12)$$

(8.3.12) says that along the optimal trajectory the rate of consumption $\dot{u}$ should be so chosen that the marginal productivity of capital as expressed by $g'(x)$ is the sum of: (1) ρ, the social rate of time prefernce; (2) δ, the rate of capital depreciation; and (3) $-(\phi''(u)/\phi'(u))\dot{u}$, the percentage rate at which the psychic cost of saving decreases over time, if we interpret $\phi'(u)$ to be the psychic cost of saving and its time rate of change as $\phi''(u)\dot{u}$. From (8.3.8) and (8.3.12) we can draw a phase diagram as shown in Figure 8.3.3. We see that $\dot{x} = 0$ if $u = g(x) - \delta x$ and $\dot{u} = 0$, if $g'(x) = \rho + \delta$. The curve

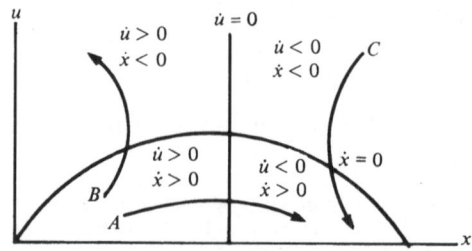

Figure 8.3.3.

$\dot{x}=0$ is drawn on the assumption that marginal productivity of capital is positive $(g'(x)>0)$ but declining $(g''(x)<0)$ and that for low levels of capital per worker $g'(x)>\delta$. Below the curve $\dot{x}=0$, consumption per capita u is less than the rate capital accumulates, hence $\dot{x}>0$ (capital per capita increases). The vertical line $\dot{u}=0$ (consumption per capita stays unchanged) is drawn at that level of x where $g'(x)=\rho+\delta$. Assuming positive but diminishing marginal utility from consumption $(\phi'>0, \phi''<0)$, we have $\dot{u}>0$ to the left of the line and $\dot{u}<0$ to the right.

The phase diagram provides us with an idea of the optimal time path that should be followed given the initial and terminal points. For example, if we follow the time path shown in the curve that was initiated at point A, it shows that both per capita consumption and capital will increase till it reaches that level of capital when $g'(x)=\rho+\delta$, after which consumption will fall but capital will continue to increase. The path that emanates from B shows that continued increasing consumption will lead to capital being eventually consumed. On the other hand, the path that emanates from C, where we start with a high level of capital, shows that continued decreasing consumption will lead to increasing capital accumulation. □

As we see from (8.3.11), the optimal control variable u derived by solving (8.3.4) or (8.3.6) is usually expressed as a function of the costate variable λ. Just as we used (8.3.10) in the above example to get the expression for $\dot{u}$ in (8.3.12), this λ may be determined by solving the canonical equations and its value then substituted into the expression for the control variable u. The following example further illustrates this situation, and is also interesting because the objective functional to be maximized will not involve an integral.

Example 3.[14] In a two-region economy, national income Y is the sum of the two regional incomes Y_1 and Y_2, $Y=Y_1+Y_2$, with $Y_i=b_ix_i$ where $x_i=$ regional capital stock and $b_i=$ constant regional output-capital ratio, $i=1,2$. National investment I is the sum of regional increases in capital stocks $\dot{x}_1$ and $\dot{x}_2$ and is equal to savings S: $I=\dot{x}_1+\dot{x}_2=S=s_1Y_1+s_2Y_2$, where $s_i=$ constant regional savings ratio. Let $g_i=s_ib_i$, $i=1,2$, be the constant regional growth rate; then

$$\dot{x}_1+\dot{x}_2=g_1x_1+g_2x_2.$$

Let $u(t)$ be the proportion of investment allocated to region 1 and $1-u(t)$ be the proportion allocated to region 2, so that

$$\dot{x}_1=u(g_1x_1+g_2x_2)=f_1(x_1,x_2,u)$$

and

$$\dot{x}_2=(1-u)(g_1x_1+g_2x_2)=f_2(x_1,x_2,u). \qquad (8.3.13)$$

---

[14]This example is from: Intriligator, M., 1964. Regional allocation of investment: comment, *Quarterly Journal of Economics*, pp. 659–662; and Rahman, Md. Anisur, 1966. Regional allocation of investment: the continuous version, *QJE*, pp. 159–60.

Assume that capital once invested in either region cannot be shifted into the other region, and assume that $u(t)$ is continuous with

$$0 \leqslant u(t) \leqslant 1. \tag{8.3.14}$$

The objective is to maximize national income $Y(T)$ at some future terminal time T (see Footnote 15) through a choice of u subject to (8.3.13) and (8.3.14) and certain given initial terminal conditions on the capital stocks x_i. That is,

$$\max_{u(t)} J = Y(T) = b_1 x_1(T) + b_2 x_2(T).$$

Notice that in this problem the intermediate function equals 0 and that the final function of (8.1.3) is

$$F(x_f, t_f) = Y(T) = b_1 x_1(T) + b_2 x_2(T).$$

We form the Hamiltonian according to (8.3.1) and (8.3.13)

$$H = \lambda_1 u(g_1 x_1 + g_2 x_2) + \lambda_2 (1 - u)(g_1 x_1 + g_2 x_2)$$
$$= [u(\lambda_1 - \lambda_2) + \lambda_2](g_1 x_1 + g_2 x_2),$$

and wish to choose that value of u which will maximize H. Clearly the optimal control $\bar{u}$ is

$$\bar{u}(t) = \begin{cases} 1 & \text{if } \lambda_1 - \lambda_2 > 0, \\ 0 & \text{if } \lambda_1 - \lambda_2 < 0, \end{cases}$$

where $0 \leqslant t < T$. We will leave it as an exercise for the reader to finish the problem.

As we pointed out, the optimal u is a function of the λ's, and it will be necessary to solve the canonical differential equations

$$\dot{\lambda}_i = -\frac{\partial H}{\partial x_i}, \qquad 0 \leqslant t < T, i = 1, 2.$$

But notice also that here we can make use of the terminal condition,

$$\lambda_i(T) = \frac{\partial Y(T)}{\partial x_i} = b_i, \qquad i = 1, 2, \tag{8.3.15}$$

to help us find the values of $\lambda_i(t)$ for the time interval $0 \leqslant t < T$.

An alternate objective for the planner of this two-region economy might be to maximize per capita consumption, instead of national income, over the time period $0 \leqslant t \leqslant T$, with total consumption C being

$$C = (1 - s_1) b_1 x_1 + (1 - s_2) b_2 x_2.$$

Assuming that population P grows at an exponential rate n and that initial population P_0 has been normalized to equal 1 as in Example 2, per capita consumption is $C \div P = e^{-nt} C$. Therefore our objective is to choose the allocation parameter $u(t)$, subject to (8.3.13) and (8.3.14), that maximizes

$$J' = \int_0^T e^{-nt} \left[(1 - s_1) b_1 x_1 + (1 - s_2) b_2 x_2 \right] dt.$$

---

[15]We use T instead of t_f here to avoid clutter.

Again using the maximum principle technique, the Hamiltonian H' is

$$H' = e^{-nt}\big[(1-s_1)b_1x_1 + (1-s_2)b_2x_2\big]$$
$$+ \lambda_1' u(g_1x_1 + g_2x_2) + \lambda_2'(1-u)(g_1x_1 + g_2x_2)$$
$$= e^{-nt}(b_1x_1 + b_2x_2) + \big[u(\lambda_1' - \lambda_2') + \lambda_2' - e^{-nt}\big](g_1x_1 + g_2x_2),$$

since $g_i = s_i b_i$. Interestingly enough, the u which yields the maximum value for H' is again

$$u(t) = \begin{cases} 1 & \text{if } \lambda_1' - \lambda_2' > 0, \\ 0 & \text{if } \lambda_1' - \lambda_2' < 0. \end{cases}$$

The reader is asked to find $\lambda_1'(t) - \lambda_2'(t)$, $0 \leqslant t \leqslant T$, by first solving the canonical equations for $\dot{\lambda}_i'$, keeping in mind that $\dot{\lambda}_i' = -\partial H'/\partial x_i$. □

8.4 Maximum principle—the costate variables and constraints

We shall not give a proof of the maximum principle.[16] Instead, we present a somewhat imprecise derivation of it to support our earlier claim in the last section that the costate variables λ_i are equivalent to the Lagrange multipliers. In the interest of simplicity, we use only one state variable $x(t)$ and one control variable $u(t)$.

In the problem

$$\max_{u(t)} J = \int_{t_0}^{t_f} I(x,u,t)\,dt + F(x_f, t_f),$$

subject to $\dot{x} = f(x,u,t)$, where u is piecewise continuous and certain boundary conditions with $x_0 = x(t_0)$ are given, we realize that the differential equation $\dot{x} = f(x,u,t)$ is a constraint of the form

$$g(x,u,t) = f(x,u,t) - \dot{x} = 0,$$

which is discussed in Subsection 8.2.3. As we said in that Subsection, this constraint must be satisfied for all t, $t_0 \leqslant t \leqslant t_f$, and we can associate with it a continuous function of time $\lambda(t)$ and treat the product $\lambda \cdot g$ under the integral by forming the following Lagrangian expression, which is a function of u, x, λ, t and $\dot{x}$:

$$L = J + \int_{t_0}^{t_f} \lambda(t)\big[f(x,u,t) - \dot{x}\big]\,dt$$
$$= \int_{t_0}^{t_f} \{I(x,u,t) + \lambda f(x,u,t) - \lambda\dot{x}\}\,dt + F(x_f, t_f)$$
$$= \int_{t_0}^{t_f} \{I + \lambda f + \dot{\lambda}x\}\,dt - \lambda x\Big|_{t_0}^{t_f} + F(x_f, t_f) \qquad (8.4.1)$$
$$= \int_{t_0}^{t_f} \{I + \lambda f + \dot{\lambda}x\}\,dt + F(x_f, t_f) - \big[\lambda(t_f)x_f - \lambda(t_0)x_0\big].$$

The last line of (8.4.1) is the result of integrating by parts (see Subsection

[16]A short, but not elementary, proof may be found in Berkovitz (1976), pp. 235–8.

8.2.1 for an earlier use of this technique) the term $\int -\lambda \dot{x}\, dt$ and then simplifying. The first two terms of the integrand is our Hamiltonian function $H = I + \lambda f$, so the Lagrangian may be written

$$L = \int_{t_0}^{t_f}(H + \dot{\lambda}x)\, dt + F(x_f, t_f) - [\lambda_f x_f - \lambda_0 x_0].$$

Assuming an interior extremum, a necessary condition for our objective functional J, hence our Lagrangian L, to attain an extremum is that $\partial L/\partial u = 0$ and $\partial L/\partial x = 0$:

$$\frac{\partial L}{\partial u} = \int_{t_0}^{t_f}\frac{\partial H}{\partial u}\, dt = 0$$

and (remembering that x_0 is fixed)

$$\frac{\partial L}{\partial x} = \int_{t_0}^{t_f}\left(\frac{\partial H}{\partial x} + \dot{\lambda}\right) dt + \frac{\partial F}{\partial x}\bigg|_{x_f} - \lambda(t_f) = 0.$$

This leads to

$$\frac{\partial H}{\partial u} = 0, \qquad \frac{\partial H}{\partial x} + \dot{\lambda} = 0, \quad \text{and} \quad \frac{\partial F}{\partial x}\bigg|_{x_f} - \lambda(t_f) = 0,$$

which are the equations given by (8.3.6), (8.3.3), and (8.3.5) respectively.

From the above we see that the costate variable λ is, with some assumptions, none other than a Lagrange multiplier. It is equivalent to the λ of the Kuhn–Tucker theorem of Chapter 7 and to the optimal value of the dual variable, y_i^o, of Chapter 6. As such, it may be thought of, as in Section 6.3, as a shadow price or accounting value.

We now look at the constraints on the control variable u. In general, it can be said that if inequality constraints are present, it usually will not be simple to incorporate the constraint analytically in a Lagrangian expression as we did with the equation of motion $\dot{x} = f(x, u, t)$. One would either use mathematical programming techniques, or (as in simple situations such as our Examples 1 and 3 in Section 8.3) consider by inspection how the Hamiltonian may be maximized by the u which satisfies the constraints, or incorporate the constraints in the control problem by imposing penalties on any violations of the constraints by u.

If the constraint on u is of integral form, say

$$\int_{t_0}^{t_f}h(u)\, dt = \text{constant},$$

then we can use the Lagrange multiplier technique again and form the Lagrangian expression

$$L = \int_{t_0}^{t_f}\{I(x, u, t) + \lambda_1[f(x, u, t) - \dot{x}] + \lambda_2 h(u)\}\, dt + F(x_f, t_f).$$

The Hamiltonian H' will be

$$H' = I(x, u, t) + \lambda_1 f(x, u, t) + \lambda_2 h(u),$$

and we will have the same set of conditions as (8.3.2)–(8.3.6), with H' replacing H. For instance, assuming that H' is maximized by an interior u,

(8.3.6) becomes

$$\frac{\partial H'}{\partial u} = \frac{\partial I}{\partial u} + \lambda_1 \frac{\partial f}{\partial u} + \lambda_2 \frac{dh}{du} = 0, \tag{8.4.2}$$

(8.3.3) becomes

$$\dot{\lambda}_1 = \frac{d\lambda_1}{dt} = -\frac{\partial H'}{\partial x} = -\left(\frac{\partial I}{\partial x} + \lambda_1 \frac{\partial f}{\partial x}\right),$$

and $\dot{\lambda}_2 = d\lambda_2/dt = -\partial H'/\partial u = 0$ from (8.4.2).

Example. Suppose we wish to maximize

$$J = \int_0^{t_f} - x^2 \, dt$$

subject to $\dot{x} = ax + bu$, $x_0 = x(0)$ given, $\int_0^{t_f} u^2 \, dt = c$, where a, b, c are constants.

The Hamiltonian H' is

$$H' = -x^2 + \lambda_1(ax + bu) + \lambda_2 u^2.$$

Using (8.4.2), we have

$$\frac{\partial H'}{\partial u} = \lambda_1 b + 2\lambda_2 u = 0, \text{ or } \bar{u} = -\frac{b}{2\lambda_2}\lambda_1.$$

Using the canonical equations given by (8.3.3) and (8.3.2), and substituting in the value of the optimal control $\bar{u}$,

$$\dot{\lambda}_1 = -\frac{\partial H'}{\partial x} = 2x - a\lambda_1,$$

$$\dot{x} = \frac{\partial H'}{\partial \lambda_1} = ax + b\bar{u} = ax - \frac{b^2}{2\lambda_2}\lambda_1.$$

These two differential equations, together with the boundary conditions $x(0) = x_0$ and $\lambda_1(t_f) = (\partial F/\partial x)|_{x_f} = 0$ (from (8.3.5); the final function $F = 0$ here), may be used to solve for the optimal control.

If there were no constraint $\int_{t_0}^{t_f} u^2 \, dt = c$ on the control variable u, the Hamiltonian for that problem would simply be:

$$H = -x^2 + \lambda(ax + bu). \qquad \square$$

8.5 Dynamic programming

Shortly before the appearance of Pontryagin's maximum principle, R. Bellman published his *Dynamic Programming*, which presents a related but different approach to the optimum design of control systems which is more efficient in some situations. A simple example will illustrate some of the main ideas behind this dynamic programming approach.

Example 1. Suppose a ship sailing from a and ending at e calls at three ports (at either of the two b's, at one of the three c's, and at one of the two d's) along the way and picks up and delivers the amounts of cargo (in hundreds of tons) as shown in Figure 8.5.1. The objective is to deliver as much cargo as possible on the entire trip. Since there are only 12 different routes, it is a simple matter to list them all and choose the route that yields the maximum tonnage. However, we shall solve the problem differently and use the following reasoning: Supposing that, somehow, we were to know the maximum tonnage values of the two shorter problems, one from b_1 to e and the other from b_2 to e, then it would be very easy to decide on the entire route. There are only two possible decisions left to be made at a: go to b_1 or go to b_2. To reach such a decision, simply add 4 to that maximum tonnage from b_1 to e that we somehow learned, add 2 to that maximum tonnage from b_2 to e, and choose the route that gives the larger value. In other words, we will have solved the original four-stage problem by first solving two three-stage problems. Similarly, each of these two three-stage problems (from b_1 to e or from b_2 to e) would be relatively easy to solve if we were to first solve three two-stage problems, namely, find the value given by the maximum tonnage path from each c_i, $i = 1, 2, 3$, to e. We continue this reasoning and reduce the process to two one-stage problems, from d_1 to e or from d_2 to e, at which stage the answer is obvious—go from d_1 because 7 is larger than 4.

Let us do the problem formally. We shall break it into several n-stage problems, $n = 1, 2, 3, 4$. Notice that there are four stages: from a to b, b to c, c to d, and d to e. There are two possible terminal ports, or *states* as we shall call them, namely b_1 and b_2 in stage one, three states c_1, c_2, and c_3 in stage two, two states d_1 and d_2 in stage three, and one state e in the last stage. Each of these states may also be thought of as the initial state for the following stages. For instance, b_1 may be considered the initial state of a three-stage problem, c_1 the initial state of a two-stage problem, *etc.* Let the variable x stand for the initial state for any n-stage problem, $n = 1, 2, 3, 4$. For instance, for a two-stage problem, x may be either c_1, or c_2, or c_3. Associated with each problem is also a *decision* or *control variable* u_n, $n = 1, 2, 3, 4$, which chooses the immediate destination when there are n stages left to go. Thus u_4 chooses b_1 or b_2, u_3 chooses c_1 or c_2 or c_3, u_2 chooses d_1 or d_2, and $u_1 = e$. Let $f_n(x, u_n)$ be the total number of tons delivered during the last n stages, given that the boat is in state x and the decision is u_n. If $\bar{u}_n$ is the decision which maximizes $f_n(x, u_n)$ for fixed n and x, let $\bar{f}_n(x)$ be that maximum value of f_n. Since $\bar{f}_n$ is the maximum value

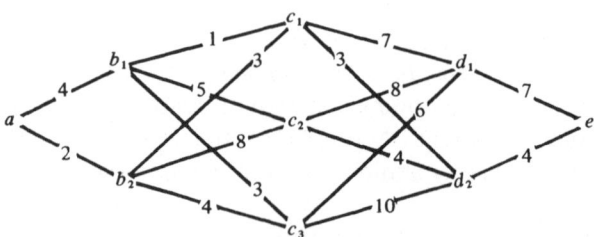

Figure 8.5.1.

Table 8.5.1   Two one-stage problems.

Problem	Initial state x	Max value $\bar{f}_1(x)$	Decision $u_1 = \bar{u}_1$
1	d_1	$7 = \bar{f}_1(d_1)$	e
2	d_2	$4 = \bar{f}_1(d_2)$	e

with respect to the decision variable u_n, it is now a function of the initial state variable x alone, hence the notation $\bar{f}_n(x)$.

In the first subproblem, there is only one stage left to go, and $\bar{u}_1 = u_1 = e$. The initial states are d_1 and d_2, as shown in Table 8.5.1.

We move now to the three subproblems in each of which there are two stages to go, but we utilize the knowledge gained from the one-stage problem. If the boat is at c_1 ($x = c_1$), it can proceed to either d_1 ($u_2 = d_1$) or d_2 ($u_2 = d_2$). If $u_2 = d_1$, $f_2(c_1, d_1) = 7 + 7 = 14$. If $u_2 = d_2$, $f_2(c_1, d_2) = 3 + 4 = 7$. Since $14 > 7$, $\bar{u}_2$ should be $\bar{u}_2 = d_1$. Similarly, if the boat is at c_2 ($x = c_2$) and $u_2 = d_1$, $f_2(c_2, d_1) = 8 + 7 = 15$, while $f_2(c_2, d_2) = 4 + 4 = 8$ if $u_2 = d_2$. Let s_{u_n} be the number of tons of cargo delivered as a result of decision u_n. Then $f_2(x, u_2) = s_{u_2} + \bar{f}_1(x)$. Table 8.5.2 shows the values for the different states and decisions.

Next we move to the two subproblems in each of which there are three stages to go, and again we utilize the knowledge gained from the previous two-stage problems. If the boat is at b_1 ($x = b_1$) and it is decided to go to c_1 ($u_3 = c_1$), the total number of tons delivered would be 1, that between b_1 and c_1, plus 14, the maximum number of tons to be delivered between c_1 and e. That is, $f_3(b_1, c_1) = s_{c_1} + \bar{f}_2(b_1)$, or $f_3(x, u_3) = s_{u_3} + \bar{f}_2(x)$. See Table 8.5.3.

The final, or four-stage, problem should now be clear. So what is the optimal policy for the overall problem? Retrace the steps backwards starting with Table 8.5.4. Starting at a, the optimal decision $\bar{u}_4$ is to go to b_2. At b_2, $\bar{u}_3$ tells us to go to c_2. At c_2, $\bar{u}_2$ tells us to go to d_1. At d_1, $\bar{u}_1$ says to

Table 8.5.2.   Three two-stage problems.

Problem	Initial state x	$f_2(x, u_2)$ $u_2 = d_1$ $f_2(x, d_1) =$ $s_{d_1} + \bar{f}_1(d_1)$	$u_2 = d_2$ $f_2(x, d_2) =$ $s_{d_2} + \bar{f}_1(d_2)$	Max value $\bar{f}_2(x)$	Optimal decision $\bar{u}_2$
1	c_1	$7 + 7 = 14$	$3 + 4 = 7$	$14 = \bar{f}_2(c_1)$	d_1
2	c_2	$8 + 7 = 15$	$4 + 4 = 8$	$15 = \bar{f}_2(c_2)$	d_1
3	c_3	$6 + 7 = 13$	$10 + 4 = 14$	$14 = \bar{f}_2(c_3)$	d_2

Table 8.5.3. Two three-stage problems.

Problem	x	$u_3 = c_1$ $f_3(x, c_1) =$ $s_{c_1} + \bar{f}_2(c_1)$	$u_3 = c_2$ $f_3(x, c_2) =$ $s_{c_2} + \bar{f}_2(c_2)$	$u_3 = c_3$ $f_3(x, c_3) =$ $s_{c_3} + \bar{f}_2(c_3)$	$\bar{f}_3(x)$	$\bar{u}_3$
			$\bar{f}_3(x, u_3)$			
1	b_1	$1 + 14 = 15$	$5 + 15 = 20$	$3 + 14 = 17$	$20 = \bar{f}_3(b_1)$	c_2
2	b_2	$5 + 14 = 19$	$8 + 15 = 23$	$4 + 14 = 18$	$23 = \bar{f}_3(b_2)$	c_2

go to e. Thus the optimal route is $a \to b_2 \to c_2 \to d_1 \to e$, with a maximum tonnage of 25.

There are only four stages in the above example and each stage has very few states, so that the computational advantages of the dynamic programming approach over the direct, brute approach of listing all twelve possible routes may not be apparent. If a problem has many stages with many states, thus involving many decision processes, direct enumeration may require a phenomenal amount of work and the computational savings of the dynamic programming approach are considerable. It has been shown that for a 20-stage problem with only 2 states in each stage, direct enumeration generates more than 1,000,000 additions, while dynamic programming requires only 220 additions.[17]

Let us point out the salient features of the above example.

It is a discrete multistage decision process problem in which one chooses a decision from a finite set of decisions at each of a finite number of stages or times. Originally, the problem consisted of n stages, but we reduced it to a sequence of n single stage decision processes, for each of which there is an optimal policy. These problems are joined together by a linear recurrence relation which we call a *functional equation*. For this particular example, the functional equation is:

$$f_n(x, u_n) = s_{u_n} + \bar{f}_{n-1}(x), \tag{8.5.1}$$

where

$$\bar{f}_n(x) = \max_{u_n} f_n(x, u_n), \qquad u = 1, 2, 3, 4.$$

In doing so, two basic ideas[18] are employed:

1. *Bellman's principle of optimality*: In a multistage decision process, given any current state, the remaining sequence of decisions forms an

[17]See Dreyfus (1965), pp. 2–6.
[18]First introduced by Bellman.

Table 8.5.4.   Four-stage problem.

	$f_4(x, u_4)$			
x	$u_4 = b_1$ $f_4(a, b_1) =$ $s_{b_1} + \bar{f}_3(b_1)$	$u_4 = b_2$ $f_4(a, b_2) =$ $s_{b_2} + f_3(b_2)$	$\bar{f}_4(x)$	$\bar{u}_4$
a	$4 + 20 = 24$	$2 + 23 = 25$	25	b_2

optimal policy with this given state regarded as the initial state. That is to say, whatever the first state and decision that led to this current state, all future decisions are optimal. In our example, if we found ourselves at, say, state c_1 (regardless of what decision led us there), the policy $c_1 \rightarrow d_1 \rightarrow e$ is optimal with c_1 considered as the initial state. Similarly, if we found ourselves at, say, state b_1, the policy $b_1 \rightarrow c_1 \rightarrow d_1 \rightarrow e$ would be optimal with b_1 considered as the initial state. By applying this principle of optimality backwards step by step repeatedly, we obtain a policy which is optimal for the overall problem. In our example, in the one-stage problems, either decision $d_1 \rightarrow e$ or $d_2 \rightarrow e$ is optimal (in fact, the only possible decision), depending on whether d_1 or d_2 is the initial state. For the two-stage problems, if c_1 is the initial state, the decision $\bar{u}_2 : c_1 \rightarrow d_1$ is optimal, and the pair $\bar{u}_2, \bar{u}_1 : c_1 \rightarrow d_1 \rightarrow e$ constitutes an optimal policy with c_1 as the initial state. If c_2 is the initial state, the pair of decisions $\bar{u}_2, \bar{u}_1 : c_2 \rightarrow d_1 \rightarrow e$ is optimal. Similarly, for the three stage problems, if b_1 is the initial state, the optimal decision $\bar{u}_3 : b_1 \rightarrow c_2$, coupled with the optimal strategy from the two-stage problem $\bar{u}_2, \bar{u}_1 : c_2 \rightarrow d_1 \rightarrow e$, form the optimal strategy $\bar{u}_3, \bar{u}_2, \bar{u}_1 : b_1 \rightarrow c_2 \rightarrow d_1 \rightarrow e$, etc.[19]

2. *The principle of imbedding*: Instead of attempting to solve a difficult problem directly, one imbeds the problem in a family of simpler, easier to solve problems and obtains the solution to the original difficult problem as a result of the solutions to the problems in the family. By repeated use of the principle of optimality, each n-stage problem with $n > 1$ is converted into a one-stage problem with its own initial state and optimal policy. This is done through the use of some functional equation such as the relation given by (8.5.1), which, for each problem in the family with its initial state, assigns an optimum value to that problem and links that value with all immediately preceding states.

These two basic ideas—imbedding and principle of optimality—are also to be found in the dynamic programming approach to continuous cases. Suppose we have a problem as stated in (8.1.3) but, for simplicity, with a

---

[19]It is not necessary to move backwards from the last stage. With some slight modifications, we can begin with the first stage and move forwards. See Dreyfus (1965), pp. 13–15.

single state variable x and single control u; that is, maximize

$$J(u) = \int_{t_0}^{t_f} I(x,u,t)\,dt + F(x_f, t_f)$$

subject to the conditions in (8.1.3). Let $\bar{J}$ be the maximum of J with respect to u. If we replace the lower limit of integration t_0 by the independent variable t, then $\bar{J}(x,t)$, where

$$\bar{J}(x,t) = \max_u \left\{ J(u,t) = \int_t^{t_f} I(x,u,t)\,dt + F(x_f, t_f) \right\} \qquad (8.5.2)$$

is the maximized value of the objective functional for a family of problems starting at time t (instead of a fixed t_0) and the corresponding initial state x (instead of the specified x_0). Hence $\bar{J}(x,t)$ is a function of x and t. The original problem is imbedded in this family of problems, and the maximized value of the objective functional for the original problem is $\bar{J}(x_0, t_0)$. We assume that the function $\bar{J}(x,t)$ has continuous first and second derivatives with respect to its arguments. The maximization process will be carried out in two steps, from t to $t+\Delta t$ and from $t+\Delta t$ to t_f. Therefore

$$\bar{J}(x,t) = \max_u \left\{ \int_t^{t+\Delta t} I(x,u,t)\,dt + \int_{t+\Delta t}^{t_f} I(x,u,t)\,dt + F(x_f, t_f) \right\}.$$

But now the principle of optimality may be invoked. Since

$$\bar{J}(x+\Delta t, t+\Delta t) = \max_u \left\{ \int_{t+\Delta t}^{t_f} I\,dt + F(x_f, t_f) \right\},$$

we have

$$\bar{J}(x,t) = \max_u \left\{ \int_t^{t+\Delta t} I\,dt + \bar{J}(x+\Delta x, t+\Delta t) \right\}. \qquad (8.5.3)$$

If we take a very short time interval Δt,

$$\int_t^{t+\Delta t} I(x,u,t)\,dt = I(x,u,t) \cdot \Delta t$$

and (8.5.3) becomes

$$\bar{J}(x,t) = \max_u \left[I\Delta t + \bar{J}(x+\Delta x, t+\Delta t) \right]. \qquad (8.5.4)$$

By Taylor's series expansion

$$\bar{J}(x+\Delta x, t+\Delta t) = \bar{J}(x,t) + \frac{\partial \bar{J}}{\partial t} \Delta t + \frac{\partial \bar{J}}{\partial x} \Delta x + \dots \quad . \qquad (8.5.5)$$

So

$$\bar{J}(x,t) = \max_u \left[I\Delta t + \bar{J}(x,t) + \frac{\partial \bar{J}}{\partial t} \Delta t + \frac{\partial \bar{J}}{\partial x} \Delta x + \dots \right]$$

or, after cancelling $\bar{J}(x,t)$ from both sides of the equation and dividing by

Δt, we have

$$0 = \max_u \left[I + \frac{\partial \bar{J}}{\partial t} + \frac{\partial \bar{J}}{\partial x} \frac{\Delta x}{\Delta t} + \cdots \right]. \tag{8.5.6}$$

As $\Delta t \to 0$, $\Delta x / \Delta t \to \dot{x} = f(x, u, t)$, and we have the following partial differential equation:

$$-\frac{\partial \bar{J}}{\partial t} = \max_u \left[I(x, u, t) + \frac{\partial \bar{J}}{\partial x} f(x, u, t) \right]. \tag{8.5.7}$$

(8.5.7) is known as *Bellman's equation*. The solution of this equation, together with the boundary conditions, would yield the solution $\bar{J}(x, t)$ and hence the particular solution $\bar{J}(x_0, t_0)$ for the original problem with its specified initial condition (x_0, t_0).

Unfortunately, the partial differential equation of (8.5.7) is usually nonlinear and possesses no analytic solution. This limitation and the restrictive assumption of continuous differentiability of $\bar{J}$ make the dynamic programming approach to continuous control problems less desirable than the maximum principle approach. It can perhaps be said that at the current level of development, dynamic programming is more promising in dealing with discrete, multistage problems.

The dynamic programming technique is applicable to a wide variety of problems, some of which (such as that of Example 1) may not involve the concept of time. In discrete time-dependent multistage problems, the time variable is usually divided into n time periods $t_1, t_2, \ldots, t_n = T$. We shall use the notation T to represent the last time period. The problem for the last time period T has x_{T-1} as its initial state, and this problem is solved first. Having found the optimal policy for this problem, we move backwards to a two-period problem consisting of the last two time periods with x_{T-2} as initial state, and so forth.

Example 2.[20] An economic planner has to make a sequence of monetary or spending policies u_t, $t = 0, 1, \ldots, T$, over a period of, say, T years which will affect the rates of unemployment x_t and inflation y_t. Suppose that these rates are related to the respective rates of the previous time period $t - 1$ by the relation

$$\begin{aligned} x_t &= a_t x_{t-1} + b_t u_t + c_t, \\ y_t &= \alpha_t y_{t-1} + \beta_t u_t + \gamma_t, \end{aligned} \qquad t = 1, \ldots, T,$$

where the $a, b, c, \alpha, \beta, \gamma$ are constants, and suppose, further, that the acceptable rate of unemployment is 4% and of inflation is 5%. The objective of the planner is to minimize

$$W = \sum_{t=1}^{T} \left\{ (x_t - 0.04)^2 + (y_t - 0.05)^2 + 2(x_t - 0.04)(y_t - 0.05) \right\}.$$

[20]This is a greatly simplified example based on a model in Chow (1975), Chapters 7 and 8.

We take x_0 and y_0 to be equal to 0. W measures the deviation of the unemployment and inflation rates from the acceptable levels. It may be considered a loss function. Employing the dynamic programming technique, we begin with the last time period T. Let

$$V_T = (x_T - 0.04)^2 + (y_T - 0.05)^2 + 2(x_T - 0.04)(y_T - 0.05)$$

$$= (a_T x_{T-1} + b_T u_T + d_T)^2 + (\alpha_T y_{T-1} + \beta_T u_T + \delta_T)^2$$

$$+ 2(a_T x_{T-1} + b_T u_T + d_T)(\alpha_T y_{T-1} + \beta_T u_T + \delta_T).$$

Notice that we allowed d_T to be $c_T - 0.04$ and δ_T to be $\gamma_T - 0.05$. It may readily be seen from below that

$$W = V_1 + \overline{V}_2.$$

We wish to minimize V_T with respect to u_T, so, assuming an interior minimum, we take the partial derivative of V_T with respect to u_T and equate that to zero to find the minimizing control $\bar{u}_T$. Then we substitute the solution $\bar{u}_T$ into V_T to find $\overline{V}_T$, the optimal, or least, loss function for time period T. After simplifications, it will be found that

$$\frac{\partial V_T}{\partial u_T} = 2(b_T + \beta_T)(a_T x_{T-1} + b_T u_T + d_T + \alpha_T y_{T-1} + \beta_T u_T + \delta_T) = 0,$$

$$\bar{u}_T = -\frac{a_T x_{T-1} + d_T + \alpha_T y_{T-1} + \delta_T}{b_T + \beta_T} = \phi(x_{T-1}, y_{T-1}),$$

$$\overline{V}_T = [a_T x_{T-1} + b_T \phi(x_{T-1}, y_{T-1}) + d_T]^2$$

$$+ [\alpha_T y_{T-1} + \beta_T \phi(x_{T-1}, y_{T-1}) + \delta_T]^2$$

$$+ 2[a_T x_{T-1} + b_T \phi(x_{T-1}, y_{T-1}) + d_T][\alpha_T y_{T-1} + \beta_T \phi(x_{T-1}, y_{T-1}) + \delta_T]$$

$$= Q_T[x_{T-1}, y_{T-1}].$$

That is, the minimizing u_T is a linear function of the states x_{T-1} and y_{T-1} of the previous time period, and the minimum loss function for the period T is a quadratic function of the states x_{T-1} and y_{T-1}.

Next, we consider the two-period problem involving the choice of two policies u_T and u_{T-1}. Since we have already found the optimal policy $\bar{u}_T$ we only need to find $\bar{u}_{T-1}$. Let

$$V_{T-1} = (x_{T-1} - 0.04)^2 + (y_{T-1} - 0.05)^2 + 2(x_{T-1} - 0.04)(y_{T-1} - 0.05) + \overline{V}_T$$

$$= (a_{T-1} x_{T-2} + b_{T-1} u_{T-1} + d_{T-1})^2 + (\alpha_{T-1} y_{T-2} + \beta_{T-1} u_{T-1} + \delta_{T-1})^2$$

$$+ 2(a_{T-1} x_{T-2} + b_{T-1} u_{T-1} + d_{T-1})$$

$$\times (\alpha_{T-1} y_{T-2} + \beta_{T-1} u_{T-1} + \delta_{T-1}) + \overline{V}_T,$$

and, proceeding as before, we find

$$\bar{u}_{T-1} = -\frac{a_{T-1}x_{T-2} + d_{T-1} + \alpha_{T-1}y_{T-2} + \delta_{T-1}}{b_{T-1} + \beta_{T-1}} = \phi(x_{T-2}, y_{T-2}),$$

$$\bar{V}_{T-1} = \left[a_{T-1}x_{T-2} + b_{T-1}\phi(x_{T-2}, y_{T-2}) + d_{T-1} \right]^2$$

$$+ \left[\alpha_{T-1}y_{T-2} + \beta_{T-1}\phi(x_{T-2}, y_{T-2}) + \delta_{T-1} \right]^2$$

$$+ 2 \left[a_{T-1}x_{T-2} + b_{T-1}\phi + d_{T-1} \right] \left[\alpha_{T-1}y_{T-1} + \beta_{T-1}\phi + \delta_{T-1} \right] + \bar{V}_T$$

$$= Q(x_{T-2}, y_{T-2}) + \bar{V}_T.$$

As we see, the optimizing decision variable $\bar{u}_{T-1}$ is the same linear function of the states of the previous time period, and the minimum loss function $\bar{V}_{T-1}$ for the two time periods T and $T-1$ is again a quadratic function of x_{T-1} and y_{T-1}. The functional relationship differs from $\bar{V}_T$ only in the addition of the last term $\bar{V}_T$ to Q.

This procedure may be repeated with the subscripts replaced by the previous time period, and so on. The optimal policy therefore is to choose the control u_t as a linear function ϕ of the state variables x_{t-1} and y_{t-1} of the preceding period. The solution W will be obtained through this process and the initial conditions $x_0 = y_0 = 0$.

8.6 Dynamic programming and the calculus of variations

We shall only give a very brief illustration of how the two approaches are not as distinct as they may appear to be. For a fuller discussion, see Dreyfus (1965).

Referring to Bellman's equation (8.5.7),

$$-\frac{\partial \bar{J}}{\partial t} = \max_u \left[I(x, u, t) + \frac{\partial \bar{J}}{\partial x} f(x, y, t) \right],$$

let $u = \dot{x}$. Then, since $\dot{x} = f(x, u, t)$, we have

$$\max_{\dot{x}} \left[I(x, \dot{x}, t) + \frac{\partial \bar{J}}{\partial x}\dot{x} + \frac{\partial \bar{J}}{\partial t} \right] = 0. \qquad (8.6.1)$$

Taking the partial derivative of each term inside the brackets with respect to $\dot{x}$ gives the necessary condition for a maximum (assuming that it occurs at an interior point $\dot{x}$):

$$\frac{\partial I}{\partial \dot{x}} + \frac{\partial \bar{J}}{\partial x} = 0. \qquad (8.6.2)$$

Substitute the maximizing $\dot{x}$ (solution to (8.6.1)) into (8.6.1) and we have

$$I + \frac{\partial \bar{J}}{\partial x}\dot{x} + \frac{\partial \bar{J}}{\partial t} = 0. \qquad (8.6.3)$$

Let us first differentiate (8.6.2) with respect to t to get

$$\frac{d}{dt}\frac{\partial I}{\partial \dot{x}} + \frac{\partial^2 \bar{J}}{\partial x \partial t}\frac{dt}{dt} + \frac{\partial^2 \bar{J}}{\partial x^2}\frac{dx}{dt} = 0, \tag{8.6.4}$$

and take the partial derivative of (8.6.3) with respect to x to get

$$\frac{\partial I}{\partial x} + \frac{\partial I}{\partial \dot{x}}\frac{\partial \dot{x}}{\partial x} + \frac{\partial^2 \bar{J}}{\partial x^2}\dot{x} + \frac{\partial \bar{J}}{\partial x}\frac{\partial \dot{x}}{\partial x} + \frac{\partial^2 \bar{J}}{\partial t \partial x} = 0,$$

which can be written

$$\frac{\partial I}{\partial x} + \left(\frac{\partial I}{\partial \dot{x}} + \frac{\partial \bar{J}}{\partial x}\right)\frac{\partial \dot{x}}{\partial x} + \frac{\partial^2 \bar{J}}{\partial x^2}\dot{x} + \frac{\partial^2 \bar{J}}{\partial t \partial x} = 0. \tag{8.6.5}$$

From (8.6.2), we know that (8.6.5) reduces to

$$\frac{\partial I}{\partial x} + \frac{\partial^2 \bar{J}}{\partial x^2}\dot{x} + \frac{\partial^2 \bar{J}}{\partial t \partial x} = 0, \quad i.e., \quad \frac{\partial^2 \bar{J}}{\partial x^2}\dot{x} + \frac{\partial^2 \bar{J}}{\partial x \partial t} = -\frac{\partial I}{\partial x}. \tag{8.6.6}$$

Substituting (8.6.6) into (8.6.4) yields

$$\frac{d}{dt}\frac{\partial I}{\partial \dot{x}} - \frac{\partial I}{\partial x} = 0,$$

which is the familiar Euler equation (8.2.9) that we have encountered before.

It can also be shown that Bellman's equation (8.5.7) implies the Hamiltonian canonical equations $\dot{x} = \partial H/\partial \lambda$ ((8.3.2)) and $\dot{\lambda} = -\partial H/\partial x$ ((8.3.3)) of the maximum principle. Recall from Section 8.4 that the Lagrangian

$$L = \int_{t_0}^{t_f}(H + \dot{\lambda}x)\,dt + F(x_f, t_f) - [\lambda_f x_f - \lambda_0 x_0].$$

At the maximum, $\bar{L} = \bar{J}$ and

$$\frac{\partial \bar{L}}{\partial x_0} = \frac{\partial \bar{J}}{\partial x_0} = \bar{\lambda}_0, \tag{8.6.7}$$

the initial value of the costate variable. In the dynamic programming approach we start each stage with its initial state, so we shall omit the subscript 0. At the maximizing u, the Hamiltonian

$$H(x, u, \lambda, t) = I(x, u, t) + \lambda f(x, u, t) = I(x, u, t) + \frac{\partial \bar{J}}{\partial x}\dot{x}, \tag{8.6.8}$$

and Bellman's equation becomes

$$-\frac{\partial \bar{J}}{\partial t} = H\left(x, u, \frac{\partial \bar{J}}{\partial x}, t\right).^{21} \tag{8.6.9}$$

---

[21]This equation is called the Hamiltonian–Jacobi equation.

Taking the partial derivative of (8.6.9) with respect to x yields

$$-\frac{\partial^2 \bar{J}}{\partial t \partial x} = \frac{\partial H}{\partial x} + \frac{\partial H}{\partial \left(\frac{\partial \bar{J}}{\partial x}\right)} \frac{\partial}{\partial x} \frac{\partial \bar{J}}{\partial x} = \frac{\partial H}{\partial x} + \frac{\partial H}{\partial \lambda} \frac{\partial^2 \bar{J}}{\partial x^2},$$

so that

$$-\frac{\partial H}{\partial x} = \frac{\partial^2 \bar{J}}{\partial t \partial x} + \frac{\partial H}{\partial \lambda} \frac{\partial^2 \bar{J}}{\partial x^2}. \tag{8.6.10}$$

Differentiating (8.6.7) with respect to t gives

$$\dot{\lambda} = \frac{d}{dt} \frac{\partial \bar{J}}{\partial x} = \frac{\partial^2 \bar{J}}{\partial x \partial t} \frac{dt}{dt} + \frac{\partial^2 \bar{J}}{\partial x^2} \frac{dx}{dt} = \frac{\partial^2 \bar{J}}{\partial x \partial t} + \frac{\partial^2 \bar{J}}{\partial x^2} \dot{x}. \tag{8.6.11}$$

From (8.6.8), $\partial H / \partial \lambda = f(x, u, t) = \dot{x}$. Therefore, comparing (8.6.10) with (8.6.11) and remembering that $\bar{J}$ is assumed to be continuously twice differentiable and that

$$\frac{\partial^2 \bar{J}}{\partial x \partial t} = \frac{\partial^2 \bar{J}}{\partial t \partial x},$$

we get the Hamiltonian canonical equations:

$$\dot{\lambda} = -\frac{\partial H}{\partial x}$$

and

$$\dot{x} = \frac{\partial H}{\partial \lambda}.$$

The main difference between the calculus of variations and dynamic programming lies in emphasis. The former considers variations of the candidate extremizing curve, whereas in dynamic programming the candidate curve varies over a small initial interval and the remainder of the curve is supposed to be optimal for the other part of the problem. In other words, the concept of variation is to be found in both approaches. Which of the two techniques is more desirable depends entirely on the needs and point of view of the user. The calculus of variations yields results whose analytical forms are useful to theorists, and its main appeal perhaps lies in solving deterministic control problems with time treated as continuous, although there are attempts to discretize time.[22] On the other hand, empiricists claim that dynamic programming is the more promising and powerful tool with wider applications in a variety of subjects. It is certainly much more efficient than the calculus of variations in dealing with stochastic control problems involving multistage decision processes.

---

[22]See, for example, Dorfman, (1969), pp. 827–8, and Benavie, Arthur, 1970. The economics of the maximum principle, *Western Economic Journal*, pp. 426–30.

8.7 Stochastic and adaptive controls

Throughout this book we have restricted ourselves to deterministic systems in which the variables are predetermined in the sense that they are not subject to random disturbances. The properties of the control processes are assumed to be completely known. In reality, of course, uncertainties and random elements are to be found in most decision processes. A growing field of study is *stochastic control*, in which the probability distributions of the random variables are entered into the formulations. The probability distributions are generally assumed to be known in this type of control. However, it often happens in some situations that this assumption is not justified. In that case the controller has to learn about the nature of certain influences which the random parameters have upon the behavior of the process. As the process unfolds and more information on the dynamic process becomes available to the controller, he adapts himself to the new situation and designs the control system accordingly. This is known as *adaptive control* and is a very promising field of study.

It is not the intent of this book to go into stochastic control or adaptive control. In fact, we have only presented an introduction to an introduction to optimal control theory, and hope to go further into the subject in another volume.

Exercises

8.1 Consider a Lagrange problem

$$\max_{\dot{x}(t)} J = \int_{t_0}^{t_1} I(x, \dot{x}, t)\, dt,$$

where x and $\dot{x}$ are $n \times 1$ vectors and the end values of x are specified. Show that by introducing a variable $\dot{x}_{n+1}(t) = I(x, \dot{x}, t)$, this problem is transformed into a Mayer problem.

8.2 (Brachistochrone problem). For the problem in which it is required to minimize

$$J = \int_{x_0}^{x_f} \sqrt{\frac{1 + (dy/dx)^2}{2g(y_0 - y)}}\, dx,$$

where g, y_0, and y_f are given, find the Euler condition, and, if desired, show that the minimizing curve γ is

$$\gamma : x = x_0 + \frac{A}{2}(\theta - \sin\theta), \qquad y = y_0 - \frac{A}{2}(1 - \cos\theta)$$

for $\theta_0 \leqslant \theta \leqslant \theta_f$, where θ_0 and θ_f are the values of θ corresponding to the endpoints $P_0 = (x_0, y_0)$, $P_f = (x_f, y_f)$, and $A = (y_0 - y)(1 + (dy/dx)^2)$.

8.3 In a macroeconomic model, we wish to minimize the sum of the squared deviations of two variables $Y(t)$ (gross national product) and $G(t)$ (government expenditure) from specified desired levels Y^* and G^*. The problem is to find optimal time paths for Y and G so as to maximize

$$-\int_{t_0}^{t_f}\left\{a(Y-Y^*)^2+c(G-G^*)^2\right\}dt$$

subject to $(1-bv)\dot{Y}+bsY-bG-bA=0$, where a, b, c, v, and s are known constants, $b=$the speed of response of supply to demand, $v=$accelerator, $s=S/Y=$savings ratio, and $A=$autonomous investment. Find the Euler–Lagrange equations and the resulting differential equations for the system.

8.4 Given $p(t)=$price of a nonrenewable resource at time t, $q(t)=$rate of output, $C(t)=C[q(t)]=$cost of extracting the resource at rate $q(t)$ with $dC/dq\geqslant0$, $R=$initial resource stock, and $r=$discount rate. A competitive producer acts to find the optimal production path to maximize the discounted present value of profit

$$\pi=\int_0^\infty e^{-rt}\left\{p(t)q(t)-C[q(t)]\right\}dt$$

subject to $\int_0^\infty q(t)\,dt\leqslant R$.[23] Find the necessary conditions for the constrained profit maximization.

8.5 A person with a known span of life T expects to earn wages at a constant rate ω and receive interest at a constant rate r on his accumulated savings or debts. Let k represent his capital, c his consumption, and α the discount rate. Then his capital accumulation is

$$\dot{k}=\omega+rk-c.$$

Assuming that $k(0)=k(T)=0$, his objective is to determine his consumption to maximize

$$\int_0^T\log ce^{-\alpha t}\,dt=\int_0^T\log(\omega+rk-\dot{k})e^{-\alpha t}\,dt.$$

Use the Euler equation and discuss his optimal consumption plan.

8.6 Do Problem 8.5 but use the maximum principle.

8.7 Read the following papers and verify the Hamiltonian functions and necessary conditions and canonical differential equations thus obtained:

(a) Blinder, Alan S., and Yoram Weiss, 1976. Human capital and labor supply: a synthesis, *Journal of Political Economy*, **84**, no. 3, pp. 449–72;

(b) Heal, Geoffrey, 1976. The relation between price and extraction cost for a resource with a backstop technology, *Bulletin of the Journal of Economics*, **7**, no. 2, pp. 371–78;

(c) Gould, John P., 1970. Diffusion processes and optimal advertising policy, *Microeconomic Foundations of Employment and Inflation Theory*, ed. Edmund S. Phelps *et al.* W. W. Norton, N.Y. In particular, pp. 348–54;

(d) Hughes, G. A., 1976. Investment and trade for a developing economy with economies of scale in industry, *Review of Economic Studies*, **43**, no. 134, pp. 237–48;

(e) Shupp, Franklin R., 1976. Optimal policy rules for a temporary income policy, *Review of Economic Studies*, **43**, no. 134, pp. 249–59.

---

[23]This topic is treated in an unpublished paper on exhaustion of nonrenewable resources by Stephen Martin, Economics Department, Michigan State University.

8.8 The following are discrete versions of the maximum principle. Compare the results with the continuous version.

(a) Dorfman (1969), pp. 827–8.
(b) Benavie, Arthur, 1970. The economics of the maximum principle, *Western Economic Journal*, **8**, no. 4, pp. 426–30.
(c) Holmes, W. L., 1968. Derivation and application of a discrete maximum principle, *Western Economic Journal*, **6**, pp. 385–94.

8.9 Refer to Example 3 of Section 8.3 and finish the problems by finding the values of $\lambda_i(t)$ and of $\lambda_1'(t) - \lambda_2'(t)$.

8.10 A man has $1,000, which he wishes to give in five installments over a period of five months to his son in amounts so that the product of the five amounts will be a maximum. Use dynamic programming to determine how much he should give each month.

8.11 A manufacturer would like to introduce a new product in three different regions A, B, and C of the country. Present estimates are that the product will not sell well in these regions with a probability 0.4, 0.5, and 0.7 respectively. Thus the present probability that the product will not sell in any region is $(0.4)(0.5)(0.7) = 0.14$. The firm has available two top salesmen that it can send to any of the three regions in order to lower this probability. The estimated probability that the product will fail in the respective regions when 0, 1, or 2 additional salesmen are sent to that region is given in the table below.

Number of additional salesmen	Region		
	A	B	C
0	0.4	0.5	0.7
1	0.2	0.3	0.45
2	0.15	0.15	0.3

Use dynamic programming to decide how many additional salesmen should be sent to which region(s). Give the functional equation.

8.12 A firm has capital K and manufactures N different products in varying quantities $q_j, j = 1, 2, \ldots, N$. The cost of producing the jth product is

$$c_j(q_j) = \begin{cases} a_j q_j + b_j & \text{if } q_j > 0, \\ 0 & \text{if } q_j = 0, \end{cases}$$

where a_j and b_j are known constants. Thus the total production cost is $\sum_{j=1}^N c_j(q_j) = K$. The profit on each unit of the jth product is p_j. Using dynamic programming, derive the functional equation for the problem of choosing q_j so as to maximize total profit $P = \sum_{j=1}^N p_j q_j$.

8.13 The work load for a certain factory varies in three periods each year according to the following estimates.

Period	I	II	III	I	II	III	
Manpower requirement	120	100	130	120	100	130	....

Employment above these levels is wasted at $5,000 per man. Bookkeeping costs of changing the level of employment from one period to another is $100 times the square of the difference in employment levels. Employment is not allowed to fall below the estimated requirement for each season. Fractional levels of employment may be interpreted as part-time work. Use dynamic programming to determine the employment level in each period in order to minimize total cost and obtain the minimum total cost for one year. *Hint*: Let u_n ($n=1,2,3$) be the employment levels for the three periods and start with Period III. The decision variable u_n is the initial state x of the following period. Hence, if r_n is the minimum manpower requirement for state n, $f_n(x,u_n)=100(u_n-x)^2+5,000(u_n-r_n)+\overline{f_{n-1}(u_n)}$.

8.14 Use dynamic programming to find Bellman's equation for the minimum time problem of Example 1, Section 8.3.

8.15 Determine the control u which minimizes

$$J(u)=\int_{t_0}^{t_f}I(x,u,t)\,dt=\int_{t_0}^{t_f}x^2\,dt$$

with $\dot{x}=-ax+bu$, a and b positive constants, x_0 given, and $|u(t)|\leqslant 1$, by employing (a) the maximum principle, (b) dynamic programming.

Appendix I: Quadratic Forms and Characteristic Roots

A *bilinear* form is

$$B = \sum_{i=1}^{m} \sum_{j=1}^{n} a_{ij} x_i y_j = a_{11} x_1 y_1 + \ldots + a_{1n} x_1 y_n$$

$$+ a_{21} x_2 y_1 + \ldots + a_{2n} x_2 y_n$$

$$\vdots$$

$$+ a_{m1} x_m y_1 + \ldots + a_{mn} x_m y_n.$$

In the special case where $m = n$ and the two sets of variables are the same, we get the *quadratic form*

$$Q(x_1, \ldots, x_n) = \sum_{i=1}^{m} \sum_{j=1}^{n} a_{ij} x_i x_j.$$

Let $A = [a_{ij}]$ be the $n \times n$ matrix of coefficients. Then Q can be written as $\mathbf{x}^T A \mathbf{x}$ where $\mathbf{x}^T = (x_1, \ldots, x_n)$. If we write $s_{ij} = \frac{1}{2}(a_{ij} + a_{ji})$, we see that $s_{ij} = s_{ji}$, and that $S = [s_{ij}]$ is a *symmetric* matrix and Q may be written as

$$Q(\mathbf{x}) = \mathbf{x}^T S \mathbf{x}. \tag{AI.1}$$

Hence we can always express a quadratic form as in (AI.1).

$Q(x_1, \ldots, x_n)$ is said to be *positive (negative) definite*, and its associated matrix S is *positive (negative) definite*, if $Q > 0$ (< 0) for any values of the variables which are not all zero. If $Q \geqslant 0$, it is said to be *positive semidefinite*.

Given any square matrix A, its characteristic equation is the determinantal equation

$$\det[A - rI] = |A - rI| = \begin{vmatrix} a_{11} - r & a_{12} & \cdots & a_{1n} \\ a_{21} & a_{22} - r & \cdots & a_{2n} \\ \vdots & & & \\ a_{n1} & a_{n2} & \cdots & a_{nn} - r \end{vmatrix} = 0, \tag{AI.2}$$

where r is a scalar and I is the identity matrix. Since (AI.2) is a polynomial equation of degree n, it will have n roots r_i, $i=1,\ldots,n$, each of which is called a *characteristic root* of the matrix A. If the matrix is symmetric, as in the case of S, its characteristic roots can be proven to be all real.

An important property of a quadratic form

$$Q(x) = \sum_{i=1}^{n} \sum_{j=1}^{n} s_{ij} x_i x_j,$$

where $S=[s_{ij}]$ is symmetric, is that there exists an orthogonal[1] matrix P such that $P^T S P$ is the diagonal matrix R of characteristic roots of S:

$$P^T S P = \text{diagonal matrix } R = [r_{ij}],$$

where $r_{ij}=0$ for $i \neq j$ and $r_{ii}=r_i$.

If we let $y=P^T x$, so that $y=P^{-1}x$ and $Py=x$, we can transform $Q(x)$ into a sum of squares:

$$Q(x) = x^T S x = y^T P^T S P y = y^T R y = \sum_{i=1}^{n} r_i y_i^2.$$

Notice that each r_i is a characteristic root of the matrix S which is associated with the quadratic form $Q(x)$. Hence we can immediately conclude that $Q(x)$ is positive definite (semidefinite) if and only if all the characteristic roots of S are positive (nonnegative), or, equivalently, if the smallest root r_s of Section 2.4 is positive (nonnegative).

As we illustrated in Section 2.4 in the case of a function of three variables, we can determine whether a function is a minimum by seeing whether the quadratic form associated with its Hessian matrix (of second-order partial derivatives) is positive definite. Hence quadratic forms play an important role in extremal problems. Closely related to the concept of quadratic forms is the concept of concavity and convexity, which we discuss briefly in Appendix II.

---

[1] A matrix P is orthogonal if $P^T = P^{-1}$, i.e., $PP^T = I$.

Appendix II: Convexity and Quasiconvexity

A set S in $\mathbf{R}^n$ is said to be a *convex set* if, for each pair of points $\mathbf{x}, \mathbf{y}$ in S, the entire line segment joining these two points is also in S. In symbols, S is convex if, given $\mathbf{x}, \mathbf{y} \in S$,

$$\mathbf{w} = k\mathbf{x} + (1-k)\mathbf{y} \Rightarrow \mathbf{w} \in S \quad \text{for } 0 \leqslant k \leqslant 1.$$

One important property of convex sets is that the intersection of any number of convex sets is also a convex set. Another property is that the collection of points which lie above (below) or on the graph of a convex (concave) function is a convex set.

This leads to the related, but different concept of convex (concave) functions.

A function $f(x_1, x_2, \ldots, x_n)$ is a *convex function* if, for each pair of points $\mathbf{x} = (x_1, \ldots, x_n)^{\mathrm{T}}$ and $\mathbf{y} = (y_1, \ldots, y_n)^{\mathrm{T}}$ on the graph of f, the line segment joining these two points lies above or on the graph of f. In symbols, f is a convex function if

$$f[k\mathbf{x} + (1-k)\mathbf{y}] \leqslant kf(\mathbf{x}) + (1-k)f(\mathbf{y}) \quad \text{for } 0 \leqslant k \leqslant 1.$$

It is *strictly convex* if $\leqslant$ is replaced by $<$ (or if the line segment joining $\mathbf{x}$ and $\mathbf{y}$ lies entirely above the graph of f except at the endpoints of the line segment).

A function f is *concave* (*strictly concave*) if we replace the word "above" by "below" or replace $\leqslant$ ($<$) by $\geqslant$ ($>$).

The second derivative, if it exists everywhere in an interval, is used to check the convexity (concavity) of a function of a single variable in that interval. Thus $f(x)$ is convex (concave) in an interval I if and only if $f''(x) \geqslant 0$ ($\leqslant 0$) for every $x \in I$. Similarly, a function of two variables $f(x_1, x_2)$ is *convex* if and only if the partial derivatives exist, and

$$\text{(i)} \quad \frac{\partial^2 f}{\partial x_1^2} \frac{\partial^2 f}{\partial x_2^2} - \left(\frac{\partial^2 f}{\partial x_1 \partial x_2} \right)^2 \geqslant 0,$$

$$\text{(ii)} \quad \frac{\partial^2 f}{\partial x_i^2} \geqslant 0 \quad \text{for } i = 1, 2.$$

$f(x_1, x_2)$ is *concave* if $\geqslant$ is replaced by $\leqslant$ in condition (ii), and it is *strictly convex* (*strictly concave*) when $\geqslant$ ($\leqslant$) is replaced by $>$ ($<$).

A function of n variables, $f(x_1, \ldots, x_n)$, is *convex* (*concave*) if and only if its Hessian matrix, or the quadratic form associated with it, is positive (negative) semidefinite. These are exactly the conditions stated in Section 2.4. For *strict convexity* (*strict concavity*), change semidefinite to definite.

The negative of a convex (concave) function is concave (convex). The sum of convex (concave) functions is convex (concave).

For a differentiable function $f(\mathbf{x})$, we can also define $f(\mathbf{x})$ to be convex if, given $\bar{\mathbf{x}}$ and any other point $\mathbf{x}$ in the domain, we have

$$f(\mathbf{x}) \geqslant f(\bar{\mathbf{x}}) + f'(\bar{\mathbf{x}})(\mathbf{x} - \bar{\mathbf{x}}).$$

Again we reverse the (strict or nonstrict) inequality for (strict or nonstrict) concavity.

In certain extremal problems, convexity may be relaxed to quasiconvexity conditions. One well-known example is the Arrow–Enthoven quasiconcave sufficiency theorem in nonlinear programming.

A function f is *quasiconvex* if, given any two distinct points $\mathbf{x}$ and $\mathbf{y}$ in the domain, we have

$$f(\mathbf{x}) \leqslant f(\mathbf{y}) \Rightarrow f[k\mathbf{x} + (1-k)\mathbf{y}] \leqslant f(\mathbf{y}) \quad \text{for } 0 < k < 1,$$

or equivalently,

$$f[k\mathbf{x} + (1-k)\mathbf{y}] \leqslant \max[f(\mathbf{x}), f(\mathbf{y})],$$

where $\max[a, b]$ means the larger of the two.

f is *quasiconcave* if

$$f(\mathbf{x}) \leqslant f(\mathbf{y}) \Rightarrow f[k\mathbf{x} + (1-k)\mathbf{y}] \geqslant f(\mathbf{x})$$

or equivalently,

$$f[k\mathbf{x} + (1-k)\mathbf{y}] \geqslant \min[f(\mathbf{x}), f(\mathbf{y})].$$

A convex (concave) function is necessarily quasiconvex (quasiconcave), but not the other way round.

Bibliography

The following is a partial list of books and papers which have been referred to in the text or may be helpful supplemental reading.

Abadie, J. (ed.), 1967. *Nonlinear Programming*. North–Holland, Amsterdam.

Allen, R. G. D., 1938. *Mathematical Analysis for Economists*. MacMillan, N.Y.

Aoki, Masanao, 1971. *Introduction to Optimization: Fundamentals and Applications of Nonlinear Programming*. MacMillan, N.Y.

Arrow, Kenneth, 1968. Applications of control theory to economic growth, *Mathematics of the Decision Sciences, Part II: Lectures in Applied Mathematics*, **12**, ed. G. B. Dantzig and A. F. Veinott. American Mathematical Society, Providence.

Barankin, E. W., and R. Dorfman, 1958. On quadratic programming, *University of California Publications in Statistics*, **2**, pp. 285–318.

Baumol, W., 1972. *Economic Theory and Operations Analysis*. 3rd ed.; Prentice Hall, N.J.

Beale, E. M. L., 1955. Cycling in the dual simplex algorithm, *Naval Research Logistics Quarterly*, **2**, pp. 269–276.

Bellman, Richard, and Robert Kalaba, 1965. *Dynamic Programming and Modern Control Theory*. Academic Press, N.Y.

Benavie, Arthur, 1972. *Mathematical Techniques for Economic Analysis*. Prentice Hall, N.J.

Berkovitz, Leonard D., 1974. *Optimal Control Theory*. Springer–Verlag, New York.

Berkovitz, Leonard D., 1976. Optimal control theory, *American Mathematical Monthly*, **83**, no. 4, pp. 225–239.

Boot, John C. G., 1964. *Quadratic Programming: Algorithms, Anomalies, Applications*. North-Holland, Amsterdam; Rand McNally, Chicago.

Chaundy, T., 1935. *The Differential Calculus*, Chapter 10. Clarendon Press, Oxford.

Charnes, A., 1952. Optimality and degeneracy in linear programming, *Econometrica*, **20**, pp. 160–170.

Chiang, Alpha, 1974. *Fundamental Methods of Mathematical Economics*. 2nd ed.; McGraw Hill, N.Y.

Chow, Gregory C., 1975. *Analysis and Control of Dynamic Economic Systems*. John Wiley, N.Y.

Converse, A. O., 1970. *Optimization*. Holt, Rinehart and Winston, N.Y.

Cooper, Leon, and David Steinberg, 1970. *Introduction to Methods of Optimization*. W. B. Saunders, Philadelphia.

Courant, R. (trans. E. J. McShane). *Differential and Integral Calculus*: Vol. I, 1937 (2nd ed.); Vol. II, 1936. Interscience, N.Y.

Dantzig, G. B., 1953. Computational algorithm of the revised simplex method, *Rand Corporation Reprint* RM-1266.

Dantzig, G. B., and R. W. Cottle, 1967. Positive (semi-) definite programming, *Nonlinear Programming*, ed. J. Abadie. North–Holland, Amsterdam.

Dantzig, G. B., and A. F. Veinott, Jr. (eds.), 1968a. *Mathematics of the Decision Sciences, Part I: Lectures in Applied Mathematics*, **11**. American Mathematical Society, Providence.

Dantzig, G. B., and A. F. Veinott, Jr. (eds.), 1968b. *Mathematics of the Decision Sciences, Part II: Lectures in Applied Mathematics*, **12**. American Mathematical Society, Providence.

Dixit, A. K., 1976. *Optimization in Economic Theory*. Oxford University Press, Oxford.

Dorfman, Robert, 1969. An economic interpretation of optimal control theory, *American Economic Review*, **59**, no. 5, pp. 817–831.

Dorfman, R., P. Samuelson, and R. Solow, 1958. *Linear Programming and Economic Analysis*. McGraw Hill, N.Y.

Dorn, W. S., 1960. A duality theorem for convex programs, *IBM Journal*, **4**, pp. 407–413.

Dorn, W. S., 1961. On Lagrange multipliers and inequalities, *Journal of Operations Research*, **9**, pp. 95–104.

Dreyfus, Stuart E., 1965. *Dynamic Programming and the Calculus of Variations*. Academic Press, N.Y.

Frisch, Ragnar (with collaboration of L. Nataf), 1966. *Maxima and Minima: Theory and Economic Applications*. Rand McNally, Chicago.

Garvin, Walter W., 1960. *Introduction to Linear Programming*. McGraw Hill, N.Y.

Hadley, G., 1962. *Linear Programming*. Addison–Wesley, Reading, Mass.

Hadley, G., 1964. *Nonlinear and Dynamic Programming*. Addison–Wesley, Reading, Mass.

Hadley, G., and M. C. Kemp, 1971. *Variational Methods in Economics*. North–Holland, Amsterdam.

Hayes, Patrick, 1975. *Mathematical Methods in the Social and Managerial Sciences*. Wiley–Interscience, N.Y.

Henderson, James M., and Richard E. Quandt, 1971. *Microeconomic Theory: A Mathematical Approach*. 2nd ed.; McGraw Hill, N.Y.

Hillier, Frederick S., and Gerald J. Lieberman, 1974. *Introduction to Operations Research*. Holden–Day, San Francisco.

Intriligator, Michael D., 1971. *Mathematical Optimization and Economic Theory*. Prentice Hall, N.J.

Kalaba, R. *See* Bellman.

Kaplan, Wilfred, 1973. *Advanced Calculus*. 2nd ed.; Addison–Wesley, Reading, Mass.

Karlin, S., 1959. *Mathematical Methods and Theory in Games, Programming and Economics*, Vol. I. Addison–Wesley, Reading, Mass.

Kemp. M. C. *See* Hadley and Kemp.

Kuhn, H. W., and A. W. Tucker, 1951. Nonlinear programming, *Proceedings, 2nd Berkeley Symposium*, pp. 481–492. University of California Press, Berkeley.

Kuhn, H. W., and A. W. Tucker (eds.), 1956. *Linear Inequallities and Related Systems*. Princeton University Press, Princeton.

Lee, E. B., and L. Markus, 1967. *Foundations of Optimal Control Theory*. John Wiley, N.Y.

Lieberman, Gerald J. *See* Hillier.

Luenberger, David G., 1973. *Introduction to Linear and Nonlinear Programming*. Addison–Wesley, Reading, Mass.

Malinvaud, E., 1972. *Lectures on Microeconomic Theory*. Elsevier North–Holland, N.Y. (With Mathematical Appendix by J. C. Milleron.)

Mangasarian, O. L., 1969. *Nonlinear Programming*. McGraw Hill, N.Y.

Markus, L. *See* Lee.

Pfaffenberger, Roger C., and David A. Walker, 1976. *Mathematical Programming for Economics and Business*. Iowa State University Press.

Pontryagin, L. S., V. G. Boltyanskii, R. V. Gamkrelidze, and E. F. Mishchenko (trans. K. N. Trinogoff; ed. L. W. Neustadt), 1962. *The Mathematical Theory of Optimal Processes*. Wiley–Interscience, N.Y.

Quandt, Richard E. *See* Henderson.

Quirk, James P., 1976. *Mathematical Notes to Intermediate Microeconomics*. Science Research Associates.

Samuelson, P. *See* Dorfman, Samuelson, and Solow.

Simon, Carl P., *to appear*. Scalar and vector maximization: calculus techniques with economics applications, *Studies in Mathematical Economics*, ed. Stanley Reiter.

Smith, Donald R., 1974. *Variational Methods in Optimization*. Prentice Hall, N.J.

Solow, R. *See* Dorfman, Samuelson, and Solow.

Spivey, W. Allen, 1963. *Linear Programming: An Introduction*. MacMillan, N.Y.

Spivey, W. Allen, and R. M. Thrall, 1970. *Linear Optimization*. Holt, Rinehart and Winston, N.Y.

Steinberg, D. *See* Cooper.

Takayama, Akina, 1974. *Mathematical Economics*. The Dryden Press, Hinsdale, Ill.

Thomas, George B., Jr., 1972. *Calculus and Analytic Geometry*. 4th ed.; Addison–Wesley, Reading, Mass.

Thompson, Gerald, 1971. *Linear Programming: An Elementary Introduction*. MacMillan, N.Y.

Thrall, R. M. *See* Spivey and Thrall.

Tou, Julius T., 1964. *Modern Control Theory*. McGraw Hill, N.Y.

Tucker, A. W. *See* Kuhn.

Veinott, A. F., Jr. *See* Dantzig.

Walker, David A. *See* Pfaffenberger.

Wolfe, Philip, 1959. The simplex method for quadratic programming, *Econometrica*, **27**, pp. 382–398.

Wolfe, P., 1961. A duality theory for nonlinear programming, *Quarterly of Applied Mathematics*, **19**, pp. 239–44.

Zangwill, W. I., 1969. *Nonlinear Programming: A Unified Approach.* Prentice Hall, N.J.

Index